Communications
in Computer and Information Science 1907

Rationale
The CCIS series is devoted to the publication of proceedings of computer science conferences. Its aim is to efficiently disseminate original research results in informatics in printed and electronic form. While the focus is on publication of peer-reviewed full papers presenting mature work, inclusion of reviewed short papers reporting on work in progress is welcome, too. Besides globally relevant meetings with internationally representative program committees guaranteeing a strict peer-reviewing and paper selection process, conferences run by societies or of high regional or national relevance are also considered for publication.

Topics
The topical scope of CCIS spans the entire spectrum of informatics ranging from foundational topics in the theory of computing to information and communications science and technology and a broad variety of interdisciplinary application fields.

Information for Volume Editors and Authors
Publication in CCIS is free of charge. No royalties are paid, however, we offer registered conference participants temporary free access to the online version of the conference proceedings on SpringerLink (http://link.springer.com) by means of an http referrer from the conference website and/or a number of complimentary printed copies, as specified in the official acceptance email of the event.

CCIS proceedings can be published in time for distribution at conferences or as post-proceedings, and delivered in the form of printed books and/or electronically as USBs and/or e-content licenses for accessing proceedings at SpringerLink. Furthermore, CCIS proceedings are included in the CCIS electronic book series hosted in the SpringerLink digital library at http://link.springer.com/bookseries/7899. Conferences publishing in CCIS are allowed to use Online Conference Service (OCS) for managing the whole proceedings lifecycle (from submission and reviewing to preparing for publication) free of charge.

Publication process
The language of publication is exclusively English. Authors publishing in CCIS have to sign the Springer CCIS copyright transfer form, however, they are free to use their material published in CCIS for substantially changed, more elaborate subsequent publications elsewhere. For the preparation of the camera-ready papers/files, authors have to strictly adhere to the Springer CCIS Authors' Instructions and are strongly encouraged to use the CCIS LaTeX style files or templates.

Abstracting/Indexing
CCIS is abstracted/indexed in DBLP, Google Scholar, EI-Compendex, Mathematical Reviews, SCImago, Scopus. CCIS volumes are also submitted for the inclusion in ISI Proceedings.

How to start
To start the evaluation of your proposal for inclusion in the CCIS series, please send an e-mail to ccis@springer.com.

Sanju Tiwari · Fernando Ortiz-Rodríguez ·
Sashikala Mishra · Edlira Vakaj · Ketan Kotecha
Editors

Artificial Intelligence: Towards Sustainable Intelligence

First International Conference, AI4S 2023
Pune, India, September 4–5, 2023
Proceedings

Springer

Editors
Sanju Tiwari ⓘ
Autonomous University of Tamaulipas
Ciudad Victoria, Mexico

Fernando Ortiz-Rodríguez ⓘ
Autonomous University of Tamaulipas
Ciudad Victoria, Mexico

Sashikala Mishra ⓘ
Symbiosis Institute of Technology
Symbiosis International (Deemed University)
Pune, Maharashtra, India

Edlira Vakaj ⓘ
Birmingham City University
Birmingham, UK

Ketan Kotecha ⓘ
Symbiosis Institute of Technology
Symbiosis International (Deemed University)
Pune, Maharashtra, India

ISSN 1865-0929   ISSN 1865-0937 (electronic)
Communications in Computer and Information Science
ISBN 978-3-031-47996-0   ISBN 978-3-031-47997-7 (eBook)
https://doi.org/10.1007/978-3-031-47997-7

This Springer imprint is published by the registered company Springer Nature Switzerland AG
The registered company address is: Gewerbestrasse 11, 6330 Cham, Switzerland

Paper in this product is recyclable.

Preface

Artificial Intelligence (AI) is a term used frequently nowadays, and its impact on different areas is getting higher. AI is expected to have a positive effect on sustainability and achieving sustainable development goals (SDGs). However, to date, there are very few published studies systematically assessing the extent to which AI might impact all aspects of sustainable development.

Given the speed with which artificial intelligence is evolving and its potential to disrupt many sectors, with advancements in Big Data, Hardware, Semantic Web, Machine Learning, Smart cities technologies, weather forecasting, and emerging powerful AI Algorithms, it seems like all the pieces are coming together to make huge changes to our everyday lives. Improving the planet Earth doesn't seem as hard as it used to be because of these advancements.

AI holds great promise for building inclusive knowledge societies and helping countries reach their targets under the 2030 Agenda for Sustainable Development, but it also poses acute ethical challenges. The current choices to develop a sustainable-development-friendly AI by 2030 have the potential to unlock benefits that could go far beyond the SDGs within our century.

The conference Artificial Intelligence: Towards Sustainable Intelligence (AI4S) aims to open a discussion on trustworthy AI and related topics, and to present the most up-to-date developments around the world from researchers and practitioners. The main scientific program of the conference comprised 16 papers: 14 full research papers and 2 short research papers selected out of 72 reviewed submissions, which corresponds to an acceptance rate of 22.2%.

The General and Program Committee chairs would like to thank the many people involved in making AI4S 2023 a success. First, our thanks go to the main chairs and the 62 reviewers for ensuring a rigorous review process that, with an average of three double-blind reviews per paper, led to an excellent scientific program. The whole AI4S team is grateful to esteemed keynote speakers Eleni Mangina, University College Dublin, Amit Sheth, University of South Carolina, and David Leslie, The Alan Turing Institute, UK for their wonderful sessions. Special acknowledgements go to Symbiosis University and the kind professors Ketan Kotecha, Deepali Vora, Shruti Patil and Sashikala Mishra for hosting the event in Pune.

Further, we are thankful for the kind support of the team at Springer. We finally thank our sponsors for their vital support for this edition of AI4S 2023. The editors would like to close the preface with warm thanks to our supporting keynotes and our enthusiastic authors who made this event truly international.

September 2023

Sanju Tiwari
Fernando Ortiz-Rodríguez
Sashikala Mishra
Edlira Vakaj
Ketan Kotecha

Organization

General Chairs

Sanju Tiwari   Universidad Autónoma de Tamaulipas, Mexico
Fernando Ortiz-Rodriguez   Universidad Autónoma de Tamaulipas, Mexico
Sashikala Mishra   Symbiosis Institute of Technology Pune, India
Edlira Vakaj   Birmingham City University, UK
Ketan Kotecha   Symbiosis Institute of Technology Pune, India

Program Chairs

Manas Gaur   UMBC, USA
Sven Groppe   University of Lübeck, Germany
Marcal Mora-Cantallops   University of Alcalá, Spain
Miguel-Angel Silcilia   University of Alcalá, Spain

Local Chairs

Deepali Vora   Symbiosis Institute of Technology, Pune, India
Shruti Patil   Symbiosis Institute of Technology, Pune, India

Publicity Chairs

Fatima Zahra Amara   University of Khenchela, Algeria
Ellie Young   Common Action, USA
Vania V. Estrela   Universidade Federal Fluminense, Brazil
Patience Usoro Usip   University of Uyo, Nigeria
Shishir Shandilya   VIT Bhopal University, India
M. A. Jabbar   Vardhman Engineering College, India

Tutorial Chairs

Sonali Vyas   UPES Dehradun, India
Kusum Lata   Sharda University, India

Rajan Gupta	Analyttica Datalab, India
Valentina Janev	University of Belgrade, Serbia
Rita Zhgeib	Canadian University Dubai, UAE

Workshop Chairs

Namrata Nagpal	Amity University, India
Meenakshi Srivastava	Amity University, India
Deepali Vora	Symbiosis Institute of Technology, Pune, India
Shruti Patil	Symbiosis Institute of Technology, Pune, India
Ghanpriya Singh	National Institute of Technology, Kurukshetra, India

Special Session Chairs

Dirdi Amna	BCU, UK
Janneth Alexandra Chicaiza Espinosa	Universidad Técnica Particular de Loja, Ecuador
Sailesh Iyer	Rai University, India
Soror Sahri	Université Paris Cité, France
Farah Benamara	IRIT, France

Program Committee

Antonio De Nicola	ENEA, Italy
Antonio López-Martín	Public University of Navarra, Spain
Anand Panchbhai	Logy.AI, India
Amna Dridi	Birmingham City University, UK
Amed Leiva-Mederos	Central University of Las Villas, Cuba
Ashwani Kumar Dubey	Amity University, India
Bhushan Zope	Symbiosis Institute of Technology, Pune, India
Carlos F. Enguix	ACM Professional Member, Peru
Csaba Csaki	Corvinus University, Hungary
Daniel Asuquo	University of Uyo, Nigeria
David Martín-Moncunill	Universidad Camilo José Cela, Spain
Deepali Vora	Symbiosis Institute of Technology, Pune, India
Edgar Tello Leal	Universidad Autónoma de Tamulipas, Mexico
Edward Udo	University of Uyo, Nigeria
Fatima Zahra Amara	University of Khenchela, Algeria

Fernando Ortiz-Rodriguez	Universidad Autónoma de Tamaulipas
Gerard Deepak	Manipal Institute of Technology, India
Gerardo Haces	Universidad Autónoma de Tamaulipas, Mexico
Ghanpriya Singh	NIT Kurukshetra, India
Gustavo de Asis Costa	Federal Institute of Education, Brazil
Hugo Eduardo Camacho Cruz	Universidad Autónoma de Tamaulipas, Mexico
Jose L. Martinez-Rodriguez	Universidad Autónoma de Tamaulipas, Mexico
Jose Melchor Medina-Quintero	Universidad Autónoma de Tamaulipas, Mexico
Jude Hemanth	Karunya University, India
Susheela Hooda	Chitkara University, India
Jitendra Kumar Samriya	UPES, Dehradun, India
Janneth Alexandra Chicaiza Espinosa	UTP de Loja, Ecuador
Kusum Lata	Sharda University, India
M. A. Jabbar	Vardhman Engineering College, India
Meenakshi Srivastava	Amity University, India
Meriem Djezzar	Khenchela University, Algeria
Miguel-Angel Sicilia	University of Alcalá, Spain
Mounir Hemam	Khenchela University, Algeria
Namrata Nagpal	Amity University, India
Neeranjan Chitare	Northumbria University, UK
Ogerta Elezaj	Birmingham City University, UK
Onur Dogan	Izmir Bakircay University, Turkey
Orchid Chetia Fukan	IIIT Delhi, India
Pankaj Goswami	Amity University, India
Patience Usoro Usip	University of Uyo, Nigeria
Praveen Kumar Shukla	BBD University, India
Rajan Gupta	University of Delhi, India
Ranjeet Bidwe	Symbiosis Institute of Technology, Pune, India
Ronak Panchal	Cognizant, India
Ritu Tanwar	NIT Uttarakhand, India
Sachinandan Mohany	VIT Andhra Pradesh, India
Sanju Tiwari	Universidad Autónoma de Tamaulipas, Mexico
Sailesh Iyer	Rai University, India
Sarra Ben Abbes	CSAI, ENGIE Lab, France
Sashikala Mishra	Symbiosis Institute of Technology Pune, India
Seema Verma	DTC Greater Noida, India
Serge Sonfack	INP-Toulouse, France
Shishir Shandilya	VIT Bhopal University, India
Shruti Patil	Symbiosis Institute of Technology, Pune, India
Sihem Benkhaled	Khenchela University, Algeria
Sonali Vyas	UPES Dehradun, India

Sourav Banerjee   Kalyani Government Engineering College, India
Soror Sahri   Université Paris Cité, France
Stavros Kourmpetis   European Banking Institute, Greece
Sven Groppe   University of Lübeck, Germany
Victor Lopez Cabrera   Universidad Tecnológica de Panamá, Panama
Yusniel Hidalgo-Delgado   Universidad de las Ciencias Informáticas, Cuba

Keynote Abstracts

Data Analytics for Sustainable Global Supply Chains

Eleni Mangina

University College Dublin, Computer Science, Dublin, Ireland

In today's modern, fast paced and globally connected world, Supply Chains have undoubtedly become a critical component of several business operations and impact our lives every day. They are crucial to the functioning of almost all the industries and every product we use and consume everyday including tooth pastes, cars, clothes, oil and electricity are products of a Supply Chain. The last century has seen rapid economic liberalization and the opening up of new markets which has allowed companies to gain access to new markets and move their production and other key economic activities to countries which offer reduced government regulation, low production costs, cheap skilled labour among other factors which help it gain a competitive edge. This globalisation has led to the creation of a Global Supply Chain network which transcends borders and this brings with it the challenges of efficiency and an increased time to market which are critical to the success of any Supply Chain. The aim of this presentation is to investigate data from the Road Freight Transport Operations in Europe, find patterns in logistic operations and analyse them based on the efficiency in terms of the vehicle utilization or the degree of loading of vehicles during each journey and sustainability in terms of amount of Carbon-dioxide emissions per journey.

Building Trustworthy Neuro-Symbolic AI Systems with Explainability and Safety: Knowledge is the Key

Amit Sheth

Director of the AI Institute at the University of South Carolina, USA

"Data alone is not enough." This was the section heading in Pedro Domingos' 2012 seminal paper. I have been a believer in this and in the duality (synergistic value) of data and knowledge for a long time. In our Semantic Search engine, commercialized in 2000, we complemented machine learning classifiers with a comprehensive WorldModel^{TM} or knowledge bases (now referred to as knowledge graphs) for improved named entity and relationship extraction and semantic search. It was an early demonstration of the complementary nature of data-driven statistical learning (since replaced by neural networks) and knowledge-supported symbolic AI methods. In this talk, I want to observe three important issues about the Why, What, and How of using knowledge in neuro-symbolic AI systems to advance from NLP to NLU. While the transformer-based models have achieved tremendous success in many NLP tasks, the pure data-driven approach comes up short when we need NLU, where knowledge is key to understanding the language, as required for the explanation, safety, and ensuring adherence to decision-making processes that must be followed (e.g., in clinical diagnosis). Throughout the talk, I will use examples from the social good domains to demonstrate the need for "understanding" (for example, for safety with explanations) and why/how knowledge-infused learning offers better outcomes compared to data-driven only alternatives.

Contents

An Approach Towards Mitigation of Renewable Energy Curtailment

Poonam Dhabai[✉] and Neeraj Tiwari

Poornima University, Jaipur, India
dhabaipoonam@gmail.com

Abstract. The expression "curtailment" refers to the cutback of power production (generation curtailment) or power consumption (load curtailment), when there is too much electricity along the grid or when there is not enough power in the grid. In contrast, curtailment focuses at minimizing the stress over the grid at an instantaneous moment. In reference to renewable energy source (RES) integrated to the stabilized grid, most commonly, curtailment is associated with the reduction of in feed from RES during its peak moment. The uncertainty in availability and predictability of RES, curtailment of RES (predominantly wind and solar energy) is an upcoming serious issue as wind and solar energy development expands across the country with increasing penetration level every fiscal year. Increasing percentage of curtailment power in feed from renewable is highly inefficient adversely affecting the revenue of RES energy projects. Curtailment is a momentarily essential but exceedingly wasteful and uneconomical form of balancing the power grid. The increasing penetration of RES has placed pressure on the power distribution centers to extract the 100% available power from RES throughout the year as per must run policy, the circumstances becomes more relentless for the period of the monsoon season when the generation from wind ramps up radically. In this work, we have demonstrated an analysis and presented an algorithm from RES (here wind & solar) generation perspective on IEEE 30 bus test system to manage and mitigate the RES curtailment.

Keywords: Renewable energy · Uncertainty · Congestion Management · Curtailment · Mitigation

1 Introduction

During the last decade, attractive policies and incentives have promoted and engrossed huge investments in renewable energy sources both at central and state level. India posses an abundance of these sources and have witnessed a ramp up in penetration level of RES generation into the network. Integration of uncertain RES in to a stabilized grid imposes numerous challenges towards the grid operators. The uncertainty factor in the availability of the RES makes it a tedious task to maintain the grid transmission stability during the high variation in RES output. The high initial investment in building up the RES plant, necessitates must run condition to ensure 100% of power output is utilized through the financial year. Due to complexity between accessing uncertain RES power output and avoiding transmission congestion, pressurizes the grid operators to maintain balance between grid stability and utilization of RES output. The operators usually plump to curtail the RES power output in the sake of transmission congestion management to maintain the stability of the grid. The curtailment increases during the peak period of RES output.

The leading states for RES inclusion in India have seen notable rise in curtailment reflecting a huge loss. Presence of congestion within the system introduces variation in the congestion cost component paraphrasing to market inefficiency. For optimality between the physical constraints and market perspective to social welfare; transmission congestion management plays a chief role. It is the need of an hour for electrical utilities to outlook a solution for efficient utilization of the existing transmission lines network by monitoring and controlling the power flow patterns as building of new transmission lines seems difficult followed by many social and environmental constraints. The integration of uncertain RES into an electrical grid changes the power flow pattern which may lead the lines into congestion thereby modifying the linear sensitivity factors of the system (i.e. Generation Shift Distribution Factor, Power Transfer Distribution Factor and Line Outage Distribution Factor). The variation of these factors leads to the change in the reliability margins (Available Transfer capability, Transmission Reliability Margin) of the transmission lines. As a result of these changes the difference in LMP values is observed which reflects the congestion cost component leading to market inefficiency.

The location and size of the RES inclusion into a grid if used appropriately can mitigate congestion from the lines [1]. Different methodologies for managing the congestion within the system includes deterministic approach (Point Estimate Method, Monte Carlo Simulation), sensitivity factors-based methods, and auction based congestion management, pricing based methods, re-dispatch of schedule and willingness to pay method [2]. Cluster based congestion management highlighting the optimal rescheduling of reactive power is proposed methodology [3]. Algorithms are developed to understand the impact and behaviour of congestion on the market concentrations [4]. Integrating sporadic RES generation into a stabilized power system possibly will entail additional cost (here congestion cost) to system owing to wind intermittency [5]. Multi objective optimization tools such as Particle Swarm Optimization, Genetic Algorithm overcomes the bottlenecks of traditional methods (Point Estimate Method, Monte Carlo simulation, weight constrained Optimal Power Flow, etc.) like computational burden, efficiency, consideration of constraint variables, etc. [5–7].

The presented algorithm examines how the location of uncertain RES power plants can introduce high congestion levels into the existing grid. To relieve the congestion from the transmission lines, the RES output is curtailed resulting in the reduction of the available utilized power of the plant. A detailed analysis based on uncertainty in the availability of the RES and the location impact is presented on standard and modified IEEE 30 bus test system. The paper presents the test case modifications, assumptions followed by solution methodology presenting the algorithm to mitigate the RES curtailment; the results are discussed later on.

2 Test Case Modifications and Assumptions

2.1 Test Case and Modifications

This work uses a standard IEEE 30 bus system test case – [Modified from Attia, A.-F., Al-Turki, Y. A., & Abusorrah, A. M. (2012)] as it is generally used for power flow analysis problems. The test system has 6 conventional generators, 21 loads, 30 bus nodes and 41transmission lines. The total conventional generation is of 190 MW with maximum generation of 60.97 MW at bus number 2. The bus, load and transmission line details are presented in previous work [33, 34]. The Fig. 1 represents the standard IEEE 30 bus test system.

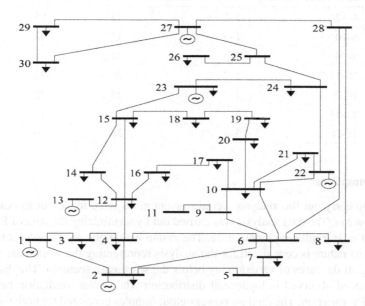

Fig. 1. Single Line Diagram of Standard IEEE 30 Bus System

The test case system is further modified and lined up in 3 different areas based on geographical parameters to analyse the congestion scenario based on variation in location of RES plants. The 3 areas are interpreted for area-based congestion management. Each area is accumulation of busses which is represented as a single bus (area). 7 tie lines connects the areas for inter area power flow transactions. The generation and tie line details are presented in Table 1 and 2 respectively.

Table 1. Area details

Parameter	Area 1	Area 2	Area3
Buses	1,2,3,4,5,6,7,8,9,11,28	12,13,14,15,16,17,18,19,20,23	10,21,22,23,24,25,26,27, 29,30
Generators	1,2	13,23	22,27
Total generation(MW)	84.5	56.2	48.5
Total load(MW)	104.5	45.1	39.6

Table 2. Tie lines details

Tie line	Interconnecting buses	Interconnecting areas	Line number	Line limit (MW)
T1	4–12	1–2	15	65
T2	6–10	1–3	12	32
T3	9–10	1–3	14	32
T4	10–17	1–2	26	32
T5	10–20	1–2	25	32
T6	23–24	2–3	32	16
T7	27–28	3–1	36	16

2.2 Assumptions

In order to carry out the analysis, certain assumptions were made out to conduct the power flow run. The data analysis was carried out by scrutinizing the data of RES wind speed and solar irradiation month wise. The month with highest variation for both RES source in its nature is considered for the analysis representing the worst-case scenario, remaining all the cases of variation lay below the worst case scenario. The distribution of wind speed observed is lognormal distribution while solar irradiation behaved in normal PDF variation. The random power output samples generated for both the sources considered are distributed normally along the mean value of data. The detailed analysis of data analysis is presented in [5–7].

The test system considered here has bus number 1 as a slack bus. The system is divided into areas on geographical base. The power flow methodology opted is DC-P-OPF. Initial transactions between the areas are done on the basis of the maximum generation and the RES farm is firstly integrated in the area with the minimum generation.

3 Solution Methodologies and Cases

3.1 Solution Methodology

The analysis is carried out in integrated fragments which includes calculation and estimation of Linear sensitivity Factors (LSF): (GSDF, PTDF & LODF), Locational Marginal Pricing (LMP), Reliability Margins (TTC & TRM). Power flow is carried out by performing inter area power transactions in MATPOWER software. The detailed calculation and estimation of above mentioned factors are presented in previous work [5–9]. The detailed algorithm for the analysis of congestion and curtailment is as presented here.

1. Obtain system parameters: bus, line, generator, load, and generator cost data, scrutinize the RES data.
2. Perform area-based power transaction depending upon the power generation and load demand.
3. Run power flow simulation to obtain base case in presence of conventional generators.
4. Record the power flow of the interconnected tie lines between the areas, calculate and estimate LSF's, and base LMP, TRM and TTC values.
5. Integrate the uncertain RES farm within the system amongst the available locations.
6. Run power flow and note the change in power flow of the transmission lines, record the new power flow.
7. Change the location of RES integration to observe the location impact and run power flow for all the available locations.
8. Calculate and estimate the LSF's, LMP's, TRM and TTC values from the new power flow in presence of wind and solar farm simultaneously for each location of RES integration.
9. Analyze the congestion scenario based on calculated factors, in presence of congestion, calculate the amount of RES to be curtailed for each location in order to minimize and mitigate the congestion.
10. Choose an optimized location for new RES farm integration based on minimized curtailment under the worst-case scenario of uncertain RES output.

The above steps are repeated for 1000 random generated samples for every month of a year for 5 different available locations of RES ($12 * 5 * 1000 = 60000$ times). The base values obtained (conventional generation only) are used as reference to compare the change occurring post RES integration. The LSF are chosen with respect to their simplicity in calculation and accuracy. GSDF indicates the generation change impact on the power flow, PTDF reflects the change in power flow of a transmission line due to change in flow of remaining transmission lines, also PTDF values are used to calculate the TTC and TRM values.

The PTDF's are used to compute the effect any transaction from Area1 to Area3 will have on any interfacing tie line. For example, for a 20-MW power transaction from Area 2 to Area 1, $20 * 0.754 = 15.08$MW of the transaction will appear on the tie line from Area2 to Area1; whereas LODF represents the contingency scenario (here N-1 contingency). The variation in TTC and TRM values are calculated and observed to analyze the presence of congestion with respect to the limits of a transmission line (voltage, thermal and stability limit). The base LMP represents the non-congestion cost, whereas the change observed in the LMP values above the base LMP values reflects addition of congestion cost. Area based congestion is analysed and managed in PSO environment to obtain an optimized location of RES farm for large generation samples and above estimated factors.

3.2   Configuration of Cases

DC-P-OPF is run firstly for base case and then for five other cases in presence of RES at different locations in the 3 areas. The details of cases are as shown in the Table 3 below.

Table 3. Case particulars

Case Number	Base Case	Case 1	Case 2	Case 3	Case 4	Case 5
Case Description	No RES	RES in Area1	RES in Area2	RES in Area3	RES in Area2	RES in Area3
Location of RES Generation	-	Bus 2	Bus 13	Bus 22	Bus 23	Bus 27

Area import and export details for conventional generation (base case) as well as wind farm (WF) and solar farm (SF) is as shown below in Table 4.

Table 4. Import and export of power between areas

CASE	Area A1(MW)		Area A2(MW)		Area A3(MW)		Import/Export (MW)		
	Generation	Load	Generation	Load	Generation	Load	A1	A2	A3
Base	84.51	104.5	56.2	45.1	48.5	39.6	–20	11.1	8.9
WF-A1	84.51 + 20	104.5	56.2	45.1	48.5	39.6	00	00	00
WF-A2	84.51	104.5	56.2 + 20	45.1	48.5	39.6	–20	20	00
WF-A3	84.51	104.5	56.2	45.1	48.5 + 20	39.6	–20	00	20
SF-A1	84.51 + 20	104.5	56.2	45.1	48.5	39.6	00	00	00
SF-A2	84.51	104.5	56.2 + 20	45.1	48.5	39.6	–20	20	00
SF-A3	84.51	104.5	56.2	45.1	48.5 + 20	39.6	–20	00	20

The power flow is run for base case and RES cases based on inter area power transactions as shown.

4   Results and Discussion

4.1   Results

Transmission lines are monopoly, the power transfer from the transmission line is restricted by the limits. Table 5, 6 and 7 interpret the results of the algorithm in the terms of PTDF values of the tie lines, percentage loading of tie lines and curtailment of RES in congestion scenario respectively, both for wind and solar farm integration simultaneously.

Table 5. PTDF Values for Tie Lines

Tie line	PTDF of Wind transmission from area 1 to area 2	PTDF of Wind transmission form area 2 to area 1
T1	0.754	0.321
T2	0.562	0.211
T3	0.048	0.012
T4	0.180	0.318
T5	0.005	0.001
T6	0.132	0.023
T7	0.413	0.212
Tie line	PTDF of Solar transmission from area 1 to area 2	PTDF of Solar transmission form area 2 to area 1
T1	0.945	0.875
T2	0.987	0.656
T3	0.787	0.984
T4	0.994	0.782
T5	1.000	0.786
T6	0.154	0.981
T7	1.000	0.812

Table 6. Area LMP and loading of lines

Case details		Area λ ($/MWh)			% Loading of Tie lines						
Case	RES Bus	Area1	Area2	Area3	T1	T2	T3	T4	T5	T6	T7
Base	-	36.35	36.35	36.35	1.4	5.6	2.8	1.2	0.9	2.4	1.1
WF-1	2	49.14	48.25	63.65	66.3	50.1	45.1	30.1	28.6	47.1	77.7
WF-2	13	45.72	51.12	45.66	60.1	58.9	58.9	63.2	67.1	51.0	51.6
	22	42.12	38.19	41.17	38.4	65.2	38.4	48.2	22.1	19.7	33.7
WF-3	23	86.75	77.78	89.84	83.1	79.2	85.6	87.4	90.1	86.1	89.1
	27	62.12	81.12	84.12	78.5	94.7	88.4	92.1	87.4	91.8	92.4
SF-1	2	36.48	44.71	42.35	28.8	17.9	22.2	18.7	38.9	31.5	25.6
SF-2	13	40.2	44.6	41.1	45.2	43.7	49.1	39.1	40.9	48.5	46.9
	22	40.2	45.9	47.9	30.8	38.5	46.8	49.1	39.2	37.3	42.3
SF-3	23	44.2	39.1	42.7	29.8	35.4	41.5	50.1	38.6	45.2	41.0
	27	37.31	38.12	37.45	15.5	22.8	30.1	19.7	19.5	16.6	14.8

Table 7. Curtailment of RES

RES Location	Bus Number	Average area LMP($/MWh)	Maximum actual power output(MW)	Curtailed power (MW)	% Curtailment
WF-A1	2	864.2	20	6.9	34.5
WF-A2	13	782.4	20	7.1	35.5
	22	682.5	20	6.4	32.0
WF-A3	23	698.5	20	9.5	47.5
	27	920.4	20	12.7	63.5
SF-A1	2	656.8	20	3.8	19.0
SF-A2	13	685.2	20	8.8	44.0
	22	656.2	20	5.2	26.0
SF-A3	23	663.8	20	6.2	31.0
	27	556.8	20	4.2	21.0

4.1.1 Base Case

For the base case (conventional only), it is observed that no or minimum congestion is present within the system. The uniformity in the LMP value in all the areas (i.e. 36.35$/MWh) reflects no addition of congestion cost. This case represents no congestion scenario with uniformity in the area price with all the generators delivering the power at the bid price. The results interpreted that the loading of the transmission lines when only conventional supplying the loads is below 5% of the TRM values. The base case area price is marked as reference value for analyzing the congestion impact.

4.1.2 Congested Case

Post integration of RES farm into the network resembled an abrupt change and variation in the power flow pattern, varying the PTDF values. The change in the PTDFs led to variation of the reliability parameters resulting to the non uniformity in the LMPs of the areas. The non uniformity reflects the occurrence of the congestion within the transmission line increasing the loading of the lines and inclusion of the congestion cost increasing the market competition. In order to balance the flow within the lines to decrease the loading and have uniformity in the LMPs, the RES is to be curtailed from its maximum output available at the given instant. The Table 5 and 6 represents the nonlinearity in the LMP values. From the results obtained, it was observed that when wind farm is located on bus 27, and solar farm on bus 13, the congestion is at its maximum with tie lines overloaded to its maximum of 95% indicating high variation in the area price values. The optimized values are obtained when wind farm is located on bus 22 and solar farm on 27 with minimum congestion and congestion cost even in the contingency case.

4.2 Discussion

The RES curtailment represents the inefficiency of the transmission grid to utilize the available power output profitably. With increasing penetration of RES in to the grid, the stabilization of the grid with maximum utilization of RES has become a major concern of the utilities. The wind power plant faces huge curtailment due to highly uncertain wind nature. The integration of new RES plants becomes more tedious if prior integrated RES are present within the grid. The location of RES plant plays an important role in managing the post congestion scenario. Irrespective of the incentives provided by government, the congestion occurrence and curtailment of RES as its solution has led the RES plants to face huge loss in terms of congestion cost. Multi objective approach is required to have an optimized solution for this problem. The results clearly indicates that inclusion of RES in to the grid leads to higher loading of transmission lines leading them to operate near or above the reliability margin values which alternately affects the market balance empowering high priced generator output rather than a healthy market competition. With this scenario, the incentives provided for promoting the inclusion of RES proves to be inefficient.

5 Conclusion and Future Scope

The aim of RES integration is to trim down pollution done by the conventional plants by increasing the involvement of power generated through them. When RES curtailment is done in the sake of transmission limits, conventional power plants linger operative, wasting the cleaner and greener electricity. This paper analyzes the affects of an individual RES plant location on system power output by investigating the transmission line congestion externality. The paper considers the location parameter of the RES plant to manage and mitigate the curtailment. The algorithm presented here shows that a social planner and grid operator taking the congestion extremities into consideration would analyze the location impact of RES thereby reducing RES curtailment. The DC-P-OPF run in the modified IEEE 30 bus test system points out that the location of a new RES plant alters the congestion levels and output at existing conventional plants. The clustered plants lead to high congestion levels lowering the total RES power output otherwise. Management and mitigation of the curtailment of power from RES is an imperative move in the direction of dipping carbon emission from conventional sources.

Acknowledgement. The authors are thankful to IMD Pune, India, for providing the necessary data. The authors are also grateful to Poornima University, Jaipur, for providing research opportunities and suitable amenities. The authors extend their thanks to Dr. A.A. Dharme, Dr. V.V. Khatavkar, and Heramb Mayadeo for their valuable guidance.

References

1. Bohn, R.E., Caramanis, M.C., Schweppe, F.C.: Optimal pricing in electrical networks over space and time. Rand J. Econ. **15**(3), 360–376 (1984). https://doi.org/10.2307/2555444

2. Christie, R.D., Wollenberg, B.F., Wangensteen, I.: Transmission management in the deregulated environment. Proc. IEEE **88**(2), 170–195 (2000). https://doi.org/10.1109/5.823997
3. Chen, L., Suzuki, H., Wachi, T., Shimura, Y.: Components of nodal prices for electric power systems. IEEE Trans. Power Syst. **17**(1), 41–49 (2002). https://doi.org/10.1109/59.982191
4. Fangxing, L.A., Bo, R.: DCPOF-based LMP simulation: algorithm, comparison with ACPF and sensitivity. In: IEEE/PES Transmission and Distribution Conference and Exposition, p. 1 (2008)
5. Dhabai, P., Tiwari, N.: Effect of stochastic nature and location change of wind and solar generation on transmission lattice power flows. In: International Conference for Emerging Technology (INCET), pp. 1–5 (2020). https://doi.org/10.1109/INCET49848.2020.9154035
6. Dhabai, P., Tiwari, N.: Analysis of variation in power flows due to uncertain solar farm power output and its location in network. In: IEEE International Conference for Innovation in Technology (INOCON), pp. 1–4 (2020). https://doi.org/10.1109/INOCON50539.2020.929 8308
7. Dhabai, P.B., Tiwari, N.: Computation of locational marginal pricing in the presence of uncertainty of solar generation. In: 5th IEEE International Conference on Recent Advances and Innovations in Engineering (ICRAIE), pp. 1–5 (2020). https://doi.org/10.1109/ICRAIE51050. 2020.9358291
8. Zhaoqiang, G., Minhui, G., Libing, Y., Dexing, W.: Locational marginal price of east china electric power market. In: 3rd International Conference on Deregulation and Restructuring and Power Technologies, DRPT 2008, pp. 214–217 (2008). https://doi.org/10.1109/DRPT. 2008.4523405
9. Prabha, S.U., Venkataseshaiah, C.: Effect of uncertainties in the economic constrained available transfer capability in power systems. Can. J. Pure Appl. Sci. **4**, 4522–4532 (2009)
10. Shayesteh, E., Parsa Moghaddam, M., Haghifam, M.R., Sheikh-EL-Eslami, M.K.: Security-based congestion management by means of demand response programs. In: IEEE Bucharest Power Tech: Innovative Ideas toward the Electrical Grid of the Future (2009). https://doi.org/ 10.1109/PTC.2009.5282069
11. Sarkar, V., Khaparde, S.A.: DCOPF-based marginal loss pricing with enhanced power flow accuracy by using matrix loss distribution. IEEE Trans. Power Syst. **24**(3), 1435–1445 (2009). https://doi.org/10.1109/TPWRS.2009.2021205
12. Kumar, A., et al.: A zonal congestion management approach using real and reactive power rescheduling. IEEE Trans. Power Syst. **19**, 554–562 (2010)
13. Ela, E., et al.: Evolution of operating reserve determination in wind power integration studies. In: IEEE PES General Meeting, PES 2010 (2010). https://doi.org/10.1109/PES.2010.5589272
14. Neuhoff, K., Boyd, R., Grau, T.: Renewable electric energy integration: quantifying the value of design of markets for international transmission capacity. IEEE Trans. **2**, 107–114 (2011)
15. Othman, M.M., Musirin, I.: A novel approach to determine transmission reliability margin using parametric bootstrap technique. Int. J. Electr. Power Energy Syst. **33**(10), 1666–1674 (2011). https://doi.org/10.1016/j.ijepes.2011.08.003
16. Ahmadi, H., Ghasemi, H.: Probabilistic optimal power flow incorporating wind power using point estimate methods. In: 10th International Conference on Environment and Electrical Engineering, EEEIC.EU Conference Proceedings (2011). https://doi.org/10.1109/EEEIC. 2011.5874815
17. Daneshi, H., Srivastava, A.K.: ERCOT electricity market: transition from zonal to nodal market operation. In: IEEE Power and Energy Society General Meeting (2011). https://doi. org/10.1109/PES.2011.6039830
18. Sjodin, E., Gayme, D.F., Topcu, U.: Risk-mitigated optimal power flow for wind powered grids. In: Proceedings of the American Control Conference, pp. 4431–4437. Institute of Electrical and Electronics Engineers Inc. (2012). https://doi.org/10.1109/acc.2012.6315377

19. Morales, J.M., Conejo, A.J., Pérez-Ruiz, J.: Simulating the impact of wind production on locational marginal prices. IEEE Trans. Power Syst. **26**(2), 820–828 (2011). https://doi.org/10.1109/TPWRS.2010.2052374
20. Moussavi, S.Z., Badri, A., Kashkooli, F.R.: Probabilistic method of wind generation placement for congestion management. World Acad. Sci. Eng. Technol. 56 (2011)
21. Attia, A.-F., Al-Turki, Y.A., Abusorrah, A.M.: Optimal power flow using adapted genetic algorithm with adjusting population size. Electr. Power Compon. Syst. **40**(11), 1285–1299 (2012). https://doi.org/10.1080/15325008.2012.689417
22. Song, Y.H., et al.: Congestion management considering voltage security constraints. In: International Conference on Power System Technology, vol. 3, pp. 13–17 (2012)
23. Banerjee, B., Jayaweera, D., Islam, S.M.: Probabilistic optimization of generation scheduling considering wind power output and stochastic line capacity. In: 22nd Australasian Universities Power Engineering Conference: Green Smart Grid Systems, AUPEC (2012)
24. Weng, Z.X., Shi, L.B., Xu, Z., Yao, L.Z., Ni, Y.X., Bazargan, M.: Effects of wind power variability and intermittency on power flow. In: IEEE Power and Energy Society General Meeting (2012). https://doi.org/10.1109/PESGM.2012.6344727
25. Lowery, C., O'Malley, M.: Optimizing wind farm locations to reduce variability and increase generation. In: International Conference on Probabilistic Methods Applied to Power Systems, PMAPS 2014 - Conference Proceedings, Institute of Electrical and Electronics Engineers Inc. (2014). https://doi.org/10.1109/PMAPS.2014.6960661
26. Bhesdadiya, R.H., Patel, R.M.: Available transfer capability calculation methods: a review. Int. J. Adv. Res. Electr. Electron. Instrum. Eng. **3**, 1–6 (2014)
27. Li, Y., Li, W., Yan, W., Yu, J., Zhao, X.: Probabilistic optimal power flow considering correlations of wind speeds following different distributions. IEEE Trans. Power Syst. **29**(4), 1847–1854 (2014). https://doi.org/10.1109/TPWRS.2013.2296505
28. Roald, L., Misra, S., Chertkov, M., Backhaus, S., Andersson, G.: Chance Constrained Optimal Power Flow with Curtailment and Reserves from Wind Power Plants. Mathematics optimization and control (2016)
29. Yu, C.-N., et al.: Congestion clusters-based markets for transmission management. In: Proceedings of IEEE PES, Winter Meeting, New York, NY, pp. 821–832 (2018)
30. Jamil, I., Zhao, J., Zhang, L., Rafique, S.F., Jamil, R.: Uncertainty analysis of energy production for a 3×50MW AC photovoltaic project based on solar resources. Int. J. Photoenergy (2019)
31. Morstyn, T., Teytelboym, A., Hepburn, C., McCulloch, M.D.: Integrating P2P energy trading with probabilistic distribution locational marginal pricing. IEEE Trans. Smart Grid **11**(4), 3095–3106 (2020). https://doi.org/10.1109/TSG.2019.2963238
32. Dhabai, P.B., Tiwari, N.: Optimization of hybrid energy generation. In: Sustainable Developments by Artificial Intelligence and Machine Learning for Renewable Energies, pp. 21–47. Elsevier (2021)
33. Dhabai, P.B., Tiwari, N.: Analysis of variation in locational marginal pricing under influence of stochastic wind generation. In: Shaw, R N., Mendis, N., Mekhilef, S., Ghosh, A. (eds.) AI and IOT in Renewable Energy. SIC, pp. 47–55. Springer, Singapore (2021). https://doi.org/10.1007/978-981-16-1011-0_6
34. Zakaryaseraji, M., Ghasemi-Marzbali, A.: Evaluating congestion management of power system considering the demand response program and distributed generation. Int. Trans. Electr. Energy Syst. **2022**, Article ID 5818757, 13 p (2022). https://doi.org/10.1155/2022/5818757

ESG and IoT: Ensuring Sustainability and Social Responsibility in the Digital Age

Federico Alberto Pozzi[1] [iD] and Dwijendra Dwivedi[2(✉)] [iD]

[1] SAS Institute, Via Carlo Darwin 20/22, 20143 Milan, Italy
[2] SAS Middle East FZ-LLC, Dubai Media City-Business Central Towers, Dubai, UAE
`dwivedy@gmail.com`

Abstract. The Internet of Things (IoT) has the potential to significantly impact Environmental, Social, and Governance (ESG) outcomes. By automating and optimizing processes and systems, IoT can help improve energy efficiency, conserve resources, and reduce pollution. It can also have social impacts, such as changing the nature of work and raising concerns about data privacy. Additionally, the governance of IoT raises important ethical and regulatory considerations. In order to ensure that the adoption of IoT contributes positively to ESG outcomes, it is important to carefully consider the potential unintended consequences and to develop and deploy the technology in a responsible and sustainable manner.

In this paper, we propose a framework based on SAS and Microsoft Azure technologies to acquire real time data from appliances, define a logic block to determine the range of data and devices to be monitored, and trigger real time alarms when needed. As the adoption of IoT continues to grow, it will be important to monitor and evaluate its impacts on ESG, and to identify and implement best practices for ensuring that IoT can contribute positively to environmental, social, and governance outcomes.

Keywords: Internet of Things · ESG · Artificial Intelligence · Industry 4.0 · SAS Intelligent Monitoring

1 Introduction

The Internet of Things (IoT) refers to the network of physical devices, vehicles, buildings, and other objects that are embedded with sensors, software, and connectivity, allowing them to collect and exchange data (Ding, 2023). The IoT allows for the automation and optimization of various tasks and processes and has the potential to transform industries and societies.

Environmental, Social, and Governance (ESG) refers to a set of criteria used to evaluate the sustainability and ethical impact of companies and organizations (Daugaard, 2020). ESG factors consider how a company performs on issues related to the environment, social responsibility, and corporate governance.

The proliferation of connected devices, collectively known as the Internet of Things (IoT), has brought about numerous benefits in terms of efficiency, productivity, and convenience. However, it has also raised concerns about environmental sustainability and

S. Tiwari et al. (Eds.): AI4S 2023, CCIS 1907, pp. 12–23, 2023.
https://doi.org/10.1007/978-3-031-47997-7_2

social responsibility. As such, the integration of environmental, social, and governance (ESG) considerations in the development and deployment of IoT is becoming increasingly important. On the environmental front, the IoT has the potential to significantly reduce resource consumption and waste through the optimization of processes and system (Farjana, 2023). For example, smart irrigation systems can adjust watering based on real-time weather data, leading to more efficient use of water resources. In transportation, connected vehicles and smart traffic management systems can reduce fuel consumption and emissions by optimizing routes and traffic flow (Agarwal, 2023).

In terms of social responsibility, the IoT has the potential to improve working conditions, safety, and the overall quality of life. For example, wearable devices can monitor the physical well-being of workers in hazardous environments and alert them to potential dangers (Chen, 2023). In the healthcare sector, connected devices can enable remote monitoring and consultations, improving access to care for underserved populations (Akkaş, 2020).

However, the integration of ESG considerations in the IoT is not without its challenges. One major issue is the potential for data privacy breaches and security vulnerabilities. As connected devices generate vast amounts of data, there is a risk that this data could be accessed by unauthorized parties or used for nefarious purposes. To address this, it is important for companies to implement robust data protection measures and ensure that their IoT products and services meet relevant privacy regulations. Another challenge is the potential for technological obsolescence and e-waste. As the IoT continues to evolve and new devices are introduced, older devices may become obsolete and end up in landfills. To mitigate this, companies can adopt a circular business model, whereby old devices are refurbished and reused instead of being discarded.

According to World Economic Forum (WEF) analysis, 84% of IoT deployments are currently addressing, or have the potential to address, Sustainable Development Goals (SDGs), which can help companies in their push for better ESG performance (World Economic Forum, 2018). IoT can be the enabler for more sustainable business and a better ESG rating in all kinds of industries: if it can be monitored, it can be measured; and if it can be measured, it can be improved.

The Sustainable Development Goals (SDGs) are a set of 17 global goals adopted by the United Nations in 2015 as part of the 2030 Agenda for Sustainable Development (Tomáš Hák, 2016). The SDGs cover a broad range of social and economic development issues, including poverty, inequality, climate change, environmental degradation, peace and justice, and partnerships for the goals. The SDGs are intended to be universal, integrated, and transformative, and to balance the economic, social, and environmental dimensions of sustainable development.

Environmental, social, and corporate governance (ESG) refers to the three central factors in measuring the sustainability and ethical impact of an investment in a company or business. These criteria help to better determine the future financial performance of companies and assist investors in decision-making. ESG criteria consider the impact of a company on the environment, its treatment of workers, customers and communities, and the way it is governed. The relationship between the SDGs and ESG is that the SDGs provide a framework for defining and measuring sustainable development, while

ESG criteria provide a framework for evaluating the sustainability and ethical impact of companies and investments (Sætra, 2021).

2 Overview of ESG and Sustainability

Environmental, social, and governance (ESG) refers to the three key areas that organizations consider when evaluating the sustainability and ethical impact of their operations. Environmental factors consider the impact of the organization on the natural environment, including issues such as greenhouse gas emissions, waste management, and resource use. Social factors consider the impact of the organization on people, including issues such as labor practices, human rights, and community engagement. Governance factors consider the internal processes and structures that govern the organization, including issues such as board composition, executive pay, and transparency. Sustainability refers to the ability of an organization to operate in a way that meets the needs of the present without compromising the ability of future generations to meet their own needs. This includes considering environmental, social, and economic factors in decision-making and seeking to balance short-term and long-term goals.

ESG and sustainability are becoming increasingly important for businesses as consumers and investors increasingly demand transparency and accountability from organizations. Companies that demonstrate a commitment to ESG and sustainability are often perceived as more trustworthy and responsible and may be more attractive to investors and customers (Krambia-Kapardis, 2023). Dwivedi (2023) found a direct link exists between an organization's ESG rating and the composition of its board of directors. Dwivedi (2023) provided a machine learning algorithm for predicting an ESG rating based on a company's financial and non-financial attributes.

2.1 Environmental, Social and Governance Impacts of IoT

One of the main ways in which IoT can impact environmental sustainability is through energy efficiency (Nitlarp, 2022). By automating and optimizing processes and systems, IoT can help reduce energy consumption and greenhouse gas emissions. For example, IoT can be used to monitor and control energy use in buildings, enabling them to be more energy efficient (Moudgil, 2023). IoT can also be used to optimize energy use in manufacturing, transportation, and other industries (Soori, 2023).

IoT can also help conserve resources by optimizing resource use and reducing waste. For example, IoT can be used to monitor and control water use in agriculture, enabling farmers to use water more efficiently. IoT can also be used to optimize resource use in manufacturing, by reducing material waste and improving supply chain efficiency. IoT can help reduce pollution by enabling the monitoring and control of emissions and waste. For example, IoT can be used to monitor and control air and water pollution, enabling companies and governments to better understand and address environmental impacts (Salman, 2023).

The adoption of IoT can also have social impacts, particularly on the workforce. IoT has the potential to change the nature of work, by automating tasks and processes that were previously performed by humans. While this may result in some job losses, it can also create new job opportunities, particularly in areas such as data analysis and machine learning.

Another social issue that is raised by IoT is data privacy. As IoT devices collect and exchange data, there are concerns about who has access to this data and how it is used. Ensuring data privacy and security will be an important consideration as the adoption of IoT continues to grow.

IoT can also have an impact on social inclusion, by providing access to information and services to underserved communities. For example, IoT can be used to provide access to healthcare, education, and other services in rural or remote areas.

The governance of IoT raises important regulatory considerations. As IoT continues to grow, there will be a need for clear and consistent regulation to ensure that the technology is used in a responsible and ethical manner.

In addition to regulatory considerations, there are also ethical considerations related to the use of IoT. For example, there are questions about the acceptable uses of data collected by IoT devices, and about the responsibility of companies and governments in ensuring the ethical use of IoT.

The integration of IoT also raises questions about corporate responsibility. Companies that adopt IoT will need to consider the environmental, social, and governance impacts of their use of the technology, and ensure that they are taking a responsible and sustainable approach.

2.2   The Contribution of Artificial Intelligence to ESG

Artificial intelligence (AI) has the potential to significantly impact and improve environmental, social, and governance (ESG) practices within organizations (Antoncic, 2020). The use of AI can improve the accuracy and efficiency of data collection, analysis, and decision-making, ultimately leading to more sustainable and responsible business practices.

One key area where AI can contribute to ESG is in the field of environmental sustainability. For example, AI algorithms can be used to optimize energy consumption and reduce carbon emissions by analyzing energy use patterns and identifying opportunities for conservation (Mhlanga, 2023). AI can also be used to monitor and predict weather patterns and climate trends, allowing organizations to make more informed decisions about their operations and how they can reduce their impact on the environment (Dewitte, 2021). AI can also contribute to social responsibility and governance within organizations. For example, AI can be used to analyze and interpret data related to diversity and inclusion within the workplace, allowing organizations to better understand and address any potential disparities or biases (Rathore, 2022). AI can also be used to monitor and analyze supply chains, helping organizations ensure that they are ethically and sustainably sourcing their materials (Toorajipour, 2021).

Overall, the use of AI can help organizations make more informed and data-driven decisions about their ESG practices, leading to more sustainable and responsible business operations. However, it is important for organizations to carefully consider the ethical implications of AI and ensure that it is being used in a responsible and transparent manner.

2.3 Industry 4.0 and Its Potential Impact on ESG

Industry 4.0, also known as the fourth industrial revolution, refers to the current trend of automation and data exchange in manufacturing technologies, including the Internet of Things (IoT), Artificial Intelligence (AI), and cloud computing (Saxena, 2022). These technologies have the potential to significantly improve the environmental, social, and governance (ESG) performance of manufacturing companies.

One way in which industry 4.0 technologies can improve ESG performance is by increasing efficiency and reducing waste. For example, the use of sensors and IoT devices can help companies monitor their production processes in real-time, allowing them to identify and address bottlenecks or inefficiencies that may be contributing to waste (Tedeschi, 2019). AI can also be used to optimize production schedules and reduce the need for raw materials, further reducing waste and environmental impact.

Other examples of industry 4.0 technologies that can improve ESG performance include (Saxena, 2022):

- Renewable energy sources, such as solar panels or wind turbines, which can be incorporated into smart factories to reduce the carbon footprint of manufacturing processes (Jebur, 2023);
- Robotics and automation, which can reduce the need for human labor, potentially improving working conditions and safety (Nyenno, 2023);
- Supply chain transparency, which can be increased through the use of blockchain and other technologies, allowing companies to track the origin and sustainability of their raw materials (Wu, 2022).

However, it is important to note that industry 4.0 technologies also have the potential to exacerbate ESG issues if not implemented carefully. For example, the automation of certain tasks may result in job loss and displacement, which can have negative social and economic impacts. Additionally, the development and deployment of new technologies may have unintended environmental consequences. It is important for companies to carefully consider the potential ESG impacts of their industry 4.0 efforts and to implement strategies to mitigate any negative impacts.

3 Proposed Approach

As SAS[1], we propose in this paper our methodology to approach ESG with IoT technology and processes. In Fig. 1 we present a schematic overview of the E2E process.

The process is represented by the following steps:

1. Appliances are connected to industrial routers and send data to them in Real Time;
2. Data are aggregated and sent to Microsoft Azure where SAS software runs;
3. A dashboard shows ESG-related KPIs (carbon footprint, energy consumed, ...) and sends alerts when a certain threshold is reached;
4. A Support Vector Data Description (SVDD) model, a machine learning method used as a one-class classifier to serve anomaly detection tasks, is developed with SAS Viya

---

[1] https://www.sas.com.

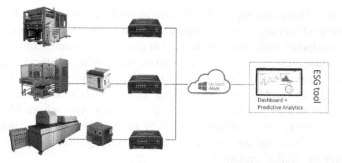

Fig. 1. Real time data collection and processing from connected appliances

Platform[2] and scored on streaming data. It has the potential to find, for example, that some industrial appliances have old engines and consumes more power than needed. Alerts are then triggered;
5. Faulty engines are replaced and the company saves energy.

3.1 Proposed Architecture

In Fig. 2 we present the reference architecture that is based on SAS and Microsoft Azure technologies. In particular, Azure will manage the security, connection and management of the devices in the cloud, makes available a platform to remotely monitor device's reliability and update configuration as needed. As SAS, we put an analytical platform and the ability to apply analytics in streaming on motion data.

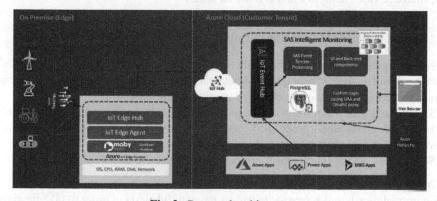

Fig. 2. Proposed architecture

Azure IoT Edge can be optionally set up as device or gateway, while IoT Hub or multiple Hubs serve as the cloud message gateway and message routing is set up to one or multiple event hubs depending on the source data, set up with built-in Apache Kafka[3] endpoint, or simply an event hub endpoint.

---

[2] https://www.sas.com/en_us/software/viya.html.
[3] https://kafka.apache.org/.

In relation to SAS technologies, SAS® Intelligent Monitoring[4], which is described in detail in the next section, uses the adapter connector as source, pointing to the event hub or Kafka endpoint (same protocol is used to get data, train and score the analytical model). This allows to bring real-time data from the real-world to Azure, using the Azure secure platform, and then launch into SAS® Intelligent Monitoring to prepare for data ingestion, acquire and profile data, define a logic block to determine the range of data to be monitored, create a project that configures an asset hierarchy, a logic block, a range of devices to monitor, alarm settings and finally run the deployment. Then the score outputs can be sent back into an Azure event hub again using direct connection to Event Hub or through the Kafka[5] endpoint.

Regarding dashboarding on the cloud side, users could use Microsoft® Power BI[6] for real-time dashboards. Power BI can inject Blob storage and can also connect to IoT hub via Azure Stream analytics, or an Azure Function. On the edge side there's the option to use SAS® Streamviewer[7] as a local dashboard to get information on data streams, events and model outputs on the edge network.

SAS can integrate with an Azure API Service, and that can be the conduit into the applications (on the right side of Fig. 2) that customers can build.

There are several benefits of using SAS and Microsoft technologies. Microsoft Azure is a Platform-as-a-Service solution that secure, connect and manage devices in the cloud. It ensures bi-directional device communication with leveraging market standard (e.g., MQTT) and is a platform to remotely monitor device's reliability and update configuration as needed. SAS has low/no code customer user experience for processing IoT Data and has a flexible, open modeling environment with Machine Learning and Deep Learning capabilities. In addition, SAS has analytics applied in streaming on motion data, image & video analytics and is built for speed.

3.2 SAS® Intelligent Monitoring: Product Overview

SAS® Intelligent monitoring is the application available from the Azure Marketplace[8] which track and analyze sensor data to detect anomalous behavior. Analysis results can be used to create rules to trigger alerts for specific conditions.

SAS® Intelligent Monitoring can be used to continually monitor the organization's assets and get alerts when predefined conditions occur. This can help with effective scheduling of maintenance and maximum asset uptime.

Specifically, SAS® Intelligent Monitoring can be used to do the following:

- Create projects targeted at assets to monitor;
- Set up analytics-based scenarios to trigger meaningful alerts in order to take the right action;

---

[4] https://documentation.sas.com/doc/en/intmoncdc/v_007/intmonug/p0cmzzg8rw8w6zn175i4e b1cgx7y.htm.

[5] https://kafka.apache.org/

[6] https://powerbi.microsoft.com/en-us/.

[7] https://documentation.sas.com/doc/en/espcdc/v_031/espvisualize/n0ydjcsczjzz3in1x2zzgtlc jsbt.htm.

[8] https://azuremarketplace.microsoft.com/en-us/.

- Deploy validated projects to monitor assets in real time;
- Modify projects to keep them running at an optimal level.

Within a Kubernetes environment, SAS® Intelligent Monitoring uses SAS® Event Stream Processing[9] to communicate with a Microsoft Azure Event Hub. The Event Hub acquires data from Microsoft IoT Hub, which is the cloud gateway for devices to publish IoT data to the Microsoft Azure cloud. With a relatively light footprint, SAS® Intelligent Monitoring uses Kubernetes services to facilitate easy management of output data. This can lead to easy consumption of alerts through downstream services such as Microsoft PowerBI.

After SAS® Intelligent Monitoring has been deployed, the workflow encapsulates a definition of assets and rules for monitoring them (Fig. 3).

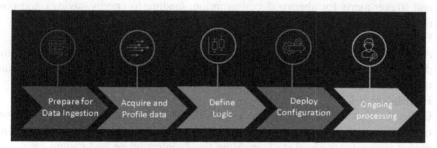

Fig. 3. The five steps of the workflow

1. Prepare for data ingestion. This involves the following:
   a. Linking the Azure IoT Hub collecting data from your device to the Azure Event Hub running in SAS® Intelligent Monitoring;
   b. Creating a JSON[10] (a language-independent data format) data structure that expresses the format of the data collected from your device;
   c. Creating input configuration files for the SAS® Intelligent Monitoring application.
2. To better understand the data that streams from your device, a data profile can be created before monitoring rules are set up. Through data injection from the Azure IoT hub, the data profile uses SAS® Event Stream Processing to acquire data from sensors on relevant assets;
3. Define a logic block to determine the range of data to be monitored;
4. Create a project that configures an asset hierarchy, a logic block, a range of devices to monitor, and alarm settings. Implement the project through a deployment. There is a one-to-one connection between a project and a deployment;
5. Run a deployment. As the deployment runs, triggered alarms are visible on the landing page. Based on what the alarms indicate, project parameters can be updated and the configuration redeployed.

---

[9] https://www.sas.com/en_us/software/event-stream-processing.html.

[10] https://www.oracle.com/database/what-is-json/.

4 Possible Applications

The proposed framework has a wide range of potential use cases across various industries. Here are some examples:

1. **Industrial IoT Monitoring**
   Use the framework to monitor and collect real-time data from sensors and equipment in manufacturing plants. Detect anomalies, predict maintenance needs, and trigger alarms for potential equipment failures to reduce downtime and improve operational efficiency.

2. **Smart Energy Management**
   Deploy the framework to monitor and analyze energy consumption patterns in buildings, factories, or campuses. Optimize energy usage, detect energy wastage, and trigger alerts for abnormal consumption, leading to energy savings and cost reduction.

3. **Predictive Maintenance in Transportation**
   Apply the framework to monitor vehicles, aircraft, or trains, gathering real-time data from onboard sensors. Predict maintenance needs, identify potential faults, and trigger alerts for proactive maintenance to ensure safety and reliability.

4. **Healthcare IoT**
   Utilize the framework to collect real-time data from medical devices and patient monitoring systems. Analyze patient vital signs, detect critical health events, and trigger alarms for medical staff, enabling timely intervention and patient care.

5. **Smart Agriculture**
   Implement the framework to monitor agricultural fields, collect data on soil moisture, temperature, and crop health. Detect anomalies, trigger irrigation systems, and notify farmers of potential issues for optimized crop yield.

6. **Connected Retail**
   Use the framework to monitor inventory levels, track product movement, and analyze customer behavior in retail stores. Trigger alarms for stock shortages, suspicious activities, or potential security breaches.

7. **Remote Asset Monitoring**
   Deploy the framework to monitor remote assets, such as oil rigs, pipelines, or weather stations. Detect equipment malfunctions, trigger alarms for hazardous conditions, and optimize maintenance schedules.

8. **Environmental Monitoring**
   Implement the framework to collect real-time data on air quality, water quality, and weather conditions. Trigger alarms for pollutant levels exceeding thresholds and monitor environmental changes.

9. **Smart City Infrastructure**
   Use the framework to monitor and manage various aspects of a smart city, such as traffic flow, waste management, and public safety. Trigger alerts for traffic congestions, waste overflow, or emergency situations.

10. **Financial Services**
    Apply the framework for real-time monitoring of financial transactions, fraud detection, and risk assessment. Trigger alarms for suspicious activities or potential security breaches.

11. **Telecommunications**

Utilize the framework to monitor network performance, track call drops, and detect unusual traffic patterns. Trigger alarms for network outages or potential cyber-attacks.

These are just a few examples of the many potential use cases for the framework. Its versatility allows it to adapt to different industries and domains, providing real-time insights and proactive alerting for critical events and operational improvements.

5 Future Work

Building on a framework that leverages SAS and Microsoft Azure technologies for real-time data acquisition, defining logic blocks, and triggering real-time alarms requires careful consideration and ongoing improvements. Here are some directions for future work to enhance the framework:

1. Data Source Integration:

   - Continuously expand the supported list of appliances and devices to cater to a wider range of IoT devices and industrial equipment;
   - Develop additional connectors and adapters to seamlessly integrate data from various sources into the framework, ensuring compatibility and easy data ingestion.

2. Data Validation and Preprocessing:

   - Enhance data validation mechanisms to ensure the integrity and reliability of incoming data;
   - Implement automatic preprocessing modules to clean, filter, and aggregate data before feeding it into the analytics engine.

3. Real-time Alarm Management:

   - Design a flexible and configurable alarm system that can trigger various actions, such as notifications, emails, or automated maintenance orders, based on severity levels;
   - Integrate Azure Notification Hubs or Event Grid to handle real-time alerts and notifications effectively.

4. Continuous Monitoring and Optimization:

   - Establish a comprehensive monitoring system to track the performance of the framework, identify bottlenecks, and optimize resource utilization.
   - Emphasize ongoing improvements through user feedback and learning from alarm triggers to fine-tune logic blocks and improve prediction accuracy.

By addressing these areas of focus, the framework based on SAS and Microsoft Azure technologies can evolve into a powerful, scalable, and efficient solution for real-time data acquisition, logic block definition, and alarm triggering across a wide range of industries and IoT applications.

6 Conclusion

In this paper, we have explored the various ways in which the Internet of Things (IoT) and Environmental, Social, and Governance (ESG) intersect and how IoT has the potential to impact ESG outcomes. Moreover, we have proposed a framework based on SAS and Microsoft Azure technologies to acquire real time data from appliances, define a logic block to determine the range of data and devices to be monitored, and trigger real time alarms when needed.

We have seen that IoT can improve environmental sustainability by reducing energy consumption and resource use, as well as reducing pollution. It can also have social impacts, such as changing the nature of work and raising concerns about data privacy. Additionally, the governance of IoT raises important ethical and regulatory considerations.

IoT has future potential in improving ESG outcomes. While IoT has the potential to greatly impact ESG outcomes, it is important to carefully consider the potential unintended consequences and ensure that the technology is developed and deployed in a responsible and sustainable manner. As the adoption of IoT continues to grow, it will be important to monitor and evaluate its impacts on ESG, and to identify and implement best practices for ensuring that IoT can contribute positively to environmental, social, and governance outcomes.

References

Agarwal, P.: Smart urban traffic management system using energy efficient optimized path discovery. In: 2023 Third International Conference on Artificial Intelligence and Smart Energy (ICAIS), pp. 858–863 (2023)

Akkaş, M.A.: Healthcare and patient monitoring using IoT. Internet Things 11, 100173 (2020)

Antoncic, M.: Uncovering hidden signals for sustainable investing using big data: artificial intelligence, machine learning and natural language processing. J. Risk Manag. Financ. Inst. 2(13), 106–113 (2020)

Chen, H.M.: The impact of wearable devices on the construction safety of building workers: a systematic review. Sustainability 15(14), 11165 (2023)

Daugaard, D.: Emerging new themes in environmental, social and governance investing: a systematic literature review. Account. Finan. 60(2), 1501–1530 (2020)

Dewitte, S.C.: Artificial intelligence revolutionises weather forecast, climate monitoring and decadal prediction. Remote Sens. 16(13), 3209 (2021)

Ding, S.T.: Opportunities and risks of internet of things (IoT) technologies for circular business models: A literature review. J. Environ. Manag. 336, 117662 (2023)

Dwivedi, D., Batra, S., Pathak, Y.K.: A machine learning based approach to identify key drivers for improving corporate's ESG ratings. J. Law Sustain. Dev. 11(1), e0242 (2023). https://doi.org/10.37497/sdgs.v11i1.242

Dwivedi, D.N., Tadoori, G., Batra, S.: Impact of women leadership and ESG ratings and in organizations: a time series segmentation study. Acad. Strateg. Manag. J. 22(S3), 1–6 (2023)

Farjana, M.F.: An IoT-and cloud-based e-waste management system for resource reclamation with a data-driven decision-making process. IoT 4(3), 202–220 (2023)

Jebur, T.K.: Greening the internet of things: a comprehensive review of sustainable IoT solutions from an educational perspective. Indones. J. Educ. Res. Technol. 3(3), 247–256 (2023)

Krambia-Kapardis, M.S.: Ethical leadership as a prerequisite for sustainable development, sustainable finance, and ESG reporting. In: Dion, M. (ed.) Sustainable Finance and Financial Crime, pp. 107–126. Springer, Cham (2023). https://doi.org/10.1007/978-3-031-28752-7_6

Mhlanga, D.: Artificial intelligence and machine learning for energy consumption and production in emerging markets: a review. Energies 16(2), 745 (2023)

Moudgil, V.H.: Integration of IoT in building energy infrastructure: a critical review on challenges and solutions. Renew. Sustain. Energy Rev. 174, 113121 (2023)

Nitlarp, T.: The impact factors of industry 4.0 on ESG in the energy sector. Sustainability 15(14), 9198 (2022)

Nyenno, I.T.: Managerial future of the artificial intelligence. Virtual Econ. 6(2), 72–88 (2023)

Rathore, B.M.: An exploratory study on role of artificial intelligence in overcoming biases to promote diversity and inclusion practices. In: Impact of Artificial Intelligence on Organizational Transformation, pp. 147–164 (2022)

Sætra, H.S.: A framework for evaluating and disclosing the ESG related impacts of AI with the SDGs. Sustainability 13, 8503 (2021)

Salman, M.Y.: Review on environmental aspects in smart city concept: water, waste, air pollution and transportation smart applications using IoT techniques. Sustain. Cities Soc. 104–567 (2023)

Saxena, A.S.: Technologies empowered environmental, social, and governance (ESG): an industry 4.0 landscape. Sustainability 1(15) (2022)

Soori, M.A.: Internet of things for smart factories in industry 4.0, a review. Internet Things Cyber-Phys. Syst. (2023)

Tedeschi, S.E.: A design approach to IoT endpoint security for production machinery monitoring. Sensors 10(19) (2019)

Tomáš Hák, S.J.: Sustainable development goals: a need for relevant indicators. Ecol. Ind. 60, 565–573 (2016)

Toorajipour, R.S.: Artificial intelligence in supply chain management: a systematic literature review. J. Bus. Res. (122), 502–517 (2021)

Wu, W., Fu, Y.: Consortium blockchain-enabled smart ESG reporting platform with token-based incentives for corporate crowdsensing. Comput. Ind. Eng. 172, 108456 (2022)

World Economic Forum. Internet of Things, Guidelines for Sustainability. World Economic Forum (2018)

AI and Assistive Technologies for Persons with Disabilities - Worldwide Trends in the Scientific Production Using Bibliometrix R Tool

Pravin Dange[1]([✉]) [iD], Tausif Mistry[2] [iD], and Shikha Mann[2] [iD]

[1] Symbiosis Institute of Management Studies, Symbiosis International (Deemed University), Pune, India
pravin.dange@sims.edu
[2] Indira School of Business Studies PGDM, Pune, India
{tausif.mistry,shikha.sindhu}@indiraisbs.ac.in

Abstract. People living with disabilities can significantly enhance their level of independence and improve their quality of life through the use of assistive technology. Recent years have seen a remarkable rise in the implementation of artificial intelligence (AI) into assistive technologies which has created novel prospects for advanced assistance and self-reliance. A comprehensive exploration of AI-based assistive technology research for individuals with disabilities is presented in this paper. Important breakthroughs are highlighted along with probable areas of further advancements. We employed bibliometrix R-tool to develop analysis while also obtaining clean metadata from Scopus database. A closer examination revealed significant emerging themes in the said topic with a clear bifurcation with basic, motor and niche themes with respect to the area of study. Interaction is a key factor in collaborations between countries and reveals the clusters emerging from the bibliometric analysis. All signs could very well point towards a forthcoming consolidation of said subject matter. In summary, it appears that there remains a need for further research in order to strengthen the field of AI-based assisted technology helping those who are disabled. While current body of literature touches on accessibility, aspects of commercialization, including afford-ability and availability, may be taken up for further research. The finding can be helpful in future research and related fields as the article provides a global view of research production over time. However, a systematic Literature review (SLR) might help in addressing the unique needs of person with disability which remain to be unaddressed by the current literature available.

Keywords: Artificial Intelligence · Person with Disability · Assistive Technology · Bibliometrix · Scopus

1 Introduction

1.1 Background

Definition of AI Based Assistive Technologies

The use of assistive technologies cannot be overlooked as it enables individuals living

S. Tiwari et al. (Eds.): AI4S 2023, CCIS 1907, pp. 24–43, 2023.
https://doi.org/10.1007/978-3-031-47997-7_3

with disabilities to manage tasks independently thereby improving their quality of life, and the use of AI-based assistive technologies is an innovative solution to tackle the specific needs and difficulties confronted by people living with disabilities. AI-assisted assistive technologies defined by Alper and Raharinirina (2006) refer to evolving devices that incorporate advanced learning models such as algorithmic intelligence along with big-data analytics for empowering persons living with disabilities.

Significance of Assistive Technologies
People with disabilities greatly benefit from assistive technologies as pointed out by Lee and Coughlin (2015), these technologies can remove barriers leading to enhanced accessibility along with fostering inclusion among individuals with disabilities. With the aid of assistive technologies that compensate or augment for disabilities enables people to attain a higher level of self-governance and independence (Kenigsberg et al., 2019). People living with physical impairments can benefit from AI-assisted mobility aids like wheelchair navigation systems which help improve their spatial awareness and enable greater freedom of movement.

Similarly, to how medical treatments are developed to enhance health outcomes for patients, assistive technologies are also developed to improve the wellbeing and general living standards for people living with disabilities. The use of the right assistive technology by people living with disabilities leads to higher levels of self-esteem as well as better social inclusion; this was evident in the analysis done by Wehmeyer and Metzler (2013). Practical support is just one benefit of these technologies; they are also instrumental in enhancing psychological well-being and granting a sense of empowerment. Additionally, access to education and employment is made easier through the incorporation of assistive technologies that reduce societal barriers while promoting economic independence (Howard et al., 2022).

1.2 Problem Statement

Challenges faced by Person with Disabilities (PwD)
According to Bartram and Cavanagh (2019), more than 1 billion people, or around 15% of the world's population, experience some form of disability. Disabilities can include physical, sensory, intellectual, and mental impairments, ranging from mobility limitations to visual or hearing impairments, cognitive disabilities, and mental health conditions. People with disabilities face many difficulties throughout their lives like impaired movement capabilities and language barriers affecting the capacity for autonomous functioning.

Several limitations exist with traditional assistive technologies that attempt to address these obstacles, and this can limit their success. However, Malhotra's argument (2004) is that traditional assistive technologies tend to be static and not adjustable based on individual needs. The outcome could be an absence of personalization coupled with difficulty in keeping up with the ever-changing requirements of individuals who have disabilities. However, those who have disabilities endure many obstacles each day - both physically and socially - which impede their ability to engage with others fully. Individuals who have disabilities are frequently confronted with obstacles when it comes to their ability to move around freely and access important information on their own according to Scherer et al. (2011). As per the research by Mankoff et al. (2010), traditional assistive technologies which cater specifically to certain disabilities or tasks can result in partial solutions that fall short of meeting all aspects of disability-related issues.

Potential of AI based Technology.
AI-based technology surpasses the limits of traditional aid and offers a broader range of solutions beyond only personal support, facilitating their use allows for progress in fields such as environmental control along with object recognition and social interaction. AI algorithms have the potential to empower assistive technologies for wheelchair navigation tasks that can significantly improve individuals' mobility and autonomy across different settings (Cortes et al. Visual impairment could be mitigated by utilizing AI-based technology to enhance the capability for recognizing objects and navigating around one's surroundings (Lee et al. Socially interactive robots powered by AI are capable enough to offer companionship and support to persons with disabilities resulting in betterment of their psychological health (Czaja and Ceruso 2022). AI-based assistive technologies have the ability to offer context-aware and highly personalized support for people with disabilities through the use of machine learning and data analytics. According to Barua et al.'s research conducted in 2022 on AI-based technologies, it was found out that these systems can master through user interaction together with data inputs. This empowers them towards adjusting functions as well as personalizing reactions thereby enhancing adaptation towards individualized requirements especially for people living with disabilities. The integration of AI-based assistive technology has promising implications for individuals with disabilities by providing comprehensive and effective support through increased learning abilities and enhanced functionality.

Developing effective AI solutions for people with blindness or vision impairment requires prioritizing HCI according to the study published in the year of 2022 by Lalanne, Baudet and Medina.

Existing Literature
Though there has been enough literature on Artificial Intelligence there isn't, however, a comprehensive overview of research into AI and assistive technologies for people with disabilities, with the exception of a few articles that are devoted to bibliographic reviews. In these articles, the authors use bibliometric techniques to analyse a variety of research articles and publications, offering insights into the research landscape, trends, and key findings in this field. In light of these factors, the purpose of this research is to identify the global trends in AI and assistive technology for people with disabilities scientific production over time.

2 Methodology

Analysis and evaluation of scientific literature can be done quantitatively through the use of bibliometric methods, which facilitates the identification of publication patterns as well as citation and collaboration within a specific field or discipline. According to Gauthier (1998) Bibliometric analysis examines bibliographic details like authorship patterns or publication date to map research output relationships between different works on a given subject, and the vital data offered by this tool enables researchers and institutions to determine scientific productivity levels accurately while also identifying impactful studies in addition to visualizing the intellectual framework of a given discipline.

Bibliometric method as per the authors Aria and Cuccurullo (2017) is definitely helpful in answering specific research questions and bibliometrics is most commonly used for answering particular research questions. This study has been developed with consideration for AI for social good ensuring reliability as well as objectivity while presenting sources, authors and documents. Authors' works across three different levels were objectively analysed through reliable sources to develop this study. The primary objective of this investigation was to recognize which topics are significant for different levels considering their productivity and citations as parameters. The analyses of knowledge structures were conducted using various bibliometric techniques, and in this study, we focused primarily on conceptual structures which outlined notable patterns in thought processes surrounding a specific topic or theme. Additionally, intellectual frameworks demonstrated how individual pieces influenced wider science communities whilst social structure showed collaboration amongst those working within the field across numerous countries. The execution of the analysis utilized a more complete range of specialized tools available with bibliometrix R Tool that are facilitated by an advanced recent R-package created by Aria and Cuccurullo (2017). If you want power and flexibility in your statistical software environment while still being able to participate through an open-source route, then R might be just what you need. As per Crawley (2007), R is an assortment of incorporated software application which are majorly used for calculating, manipulating data and displaying graphics. Bibliometrix has the capability of being integrated with additional related software, and the data collection was followed by a descriptive as well as bibliometric evaluation to conduct the performative bibliography assessment.

2.1 Data Collection

Data for bibliometric analysis research has been collected using the Scopus database, a comprehensive resource for scholarly literature. The database allows researchers to search and retrieve relevant bibliographic information, including publication metadata, citations, author affiliations, and keywords, enabling in-depth analysis of scientific publications (López-Pernas, Saqr, and Apiola (2023). The first stage of collecting data was identification of the records using Boolean operators available. This search took place on 22nd May 2023.

It is important to highlight that the concepts – "Artificial Intelligence", "Technology assisted Tools" and "Person with Disabilities" were placed in inverted commas to tell the software to consider the combined meaning of the words. The Boolean operator used was "AND" to search the database. After running the query, the database gave 249 documents. Since the year 2023 is still in progress, we applied the filer for excluding the year 2023 Research on AI-based Technology assisted tools on Person with Disability, we found limited articles on the said topic and can be concluded it is a very niche area. The researcher had narrowed down the research using the Boolean operators in the search query and the entire record including book chapters have been included in the study. The bibliometric analysis was conducted on these 225 records extracted and loaded in Bibliometrix Tool developed by Aria and Cuccurullo (2017). The step-by-step data collection can be summarized as:

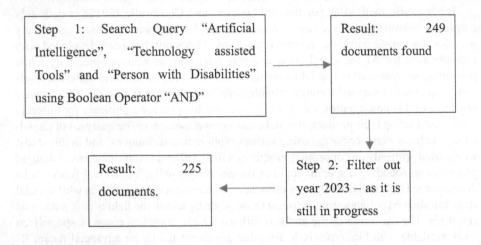

Step 1: Search Query "Artificial Intelligence", "Technology assisted Tools" and "Person with Disabilities" using Boolean Operator "AND"

Result: 249 documents found

Result: 225 documents.

Step 2: Filter out year 2023 – as it is still in progress

2.2 Analysis

For the analysis of the data, descriptive statistics of the entire data was performed. Post the descriptive statistics, bibliometric techniques were developed using conceptual and intellectual networks for each level of analysis (Table 1).

Loi M (2016) argued that by examining co-citations we could understand both direct links of the citation and the intellectual structure of research fields. Additionally, the citation frequency of a third document reveals the strength of this network. If there is significant co-citation link between two papers then it's highly probable that they will be referred to independently too. Moreover, the collaboration between authors can be characterized based on their intellect or social ties using the co-author analysis. Researchers can utilize this analysis in order to identify previously unknown research groups (Otte and Rousseau 2002). Additionally, it helps in identifying collaborations between various countries who act as collaborating pairs (Glänzel and Schubert 2006). In a co-author network diagramming creators and nations as nodes makes it simple to understand collaboration through connections. Using guidelines from Table 3 each analysis level was characterized resulting in representation of its related co-citation and collaboration network graphs. An alternative way to calculate vertex distances using random walks was proposed by Pons and Latapy (2006) through their development of the Walktrap algorithm. It possesses advantages such as capturing copious amounts of data or efficiently detecting communities within networks. Kamada and Kaway (1989) proposed an intuitive way of representing undirected graphs where vertices location reflects its relationship with other vertices. Furthermore, co-word analysis is based on the belief that when certain words occur together, they depict contents of documents in a file (Callon et al. 1991). With regard to this matter, Cobo et al.'s (2011) proposition suggested combining performance analysis and scientific mapping in order to examine one particular field of research from a conceptual perspective. To determine evolving

Table 1. Specification of Analysis

Level of Analysis	Metrics	Unit of Analysis	Bibliometric Technique	Statistical Technique	Structure
Source	Source Dynamics & Most Productive Sources	Journal	Co-citation	Network	Conceptual
Authors	Most productive authors & Annual production per author	Authors	Co-citation & Collaboration	Network	Intellectual and Social
Documents	Most productive countries	Countries	Collaboration	Network	Social
	Most Cited Documents	References	Co-citation	Network	Intellectual
	Most Frequent Author Keywords (KE) & Keyword Plus (KP)	Author Keywords (KE) & Keyword Plus (KP)	Co-words	Thematic mapping and Thematic evolution	Conceptual

thematic trends throughout the period the authors proposed employing co-word analysis to detect these distinguishable subjects and representing them via strategic charts.

In order to effectively represent the result of cooccurrence clustering, it is important to consider both centrality and density. Clusters' connectivity to each other represents their centrality while their internal cohesiveness reflects their density. In order to arrange the detected clusters of any given sub-period on a strategic diagram and start conducting dynamic analysis, one can consider using these two measures. When examining a strategic diagram, you will notice that it contains four different zones or quadrants. To specify further, the theme zones include motor focused topics which encompass bodily movements coordinated by the brain. Developed or isolated topics requiring in-depth research may also fall under these zones. Basic or Transversal subjects with applications to diverse fields represent another category also part of these theme zones along with those related to external concepts. It's probable that particular themes which had high centrality and density in Zone 1 have been researched over an extended period by a well-defined group. The appearance of clusters within Zone 2 may suggest potential areas for future research themes to emerge or provide linkages between otherwise disconnected networks. In zone 3 there exist groups of themes that have a history of development dating back to when they existed in area one although they are less important nowadays. The contribution of clusters that exist only within zone 4 can be determined solely through a careful evaluation utilizing dynamic analysis. Also, a cluster's volume depends on how many terms are inside it while its designated name depicts whichever phrase gets used most commonly in connection with those terms. The main analytical tool in use for this research was keyword plus (KP), as it provided a much larger pool of relevant terms when compared to author keywords. Like before we used the Walktrap algorithm for performing the clustering, thematic can be understood as an amalgamation of developed themes transcending various sub-periods according to the author's definition. Based on their interconnectivity with each other, certain themes might merely be independent entities without any relation to other similar clusters.

3 Results and Discussion

3.1 Production

According to the scientific production of literature from 1995 to 2022 and the main topics associated with bibliometric analysis as given in table above (Table 2), it can be seen that there has been a significant increase in the number of literatures produced. The number of articles produced till the year 2015 was a single digit number however research in the field has seen a rapid increase in the past couple of years. The number of productions almost doubled in the year 2021 as compared to the previous year and has since then been in a maturity stage. (Fig. 1).

3.2 Sources

With regards to the sources, it can be noted that *Lecture Notes in Computer Science (Including Subseries Lecture Notes in Artificial Intelligence and Lecture Notes in Bioinformatics)* was the leading source (Fig. 2).

The source had the production of scientific literature on the said topic with a total publication in 2022 being 14 articles followed by *ACM International Conference Proceeding Series* and *Disability and Rehabilitation: Assistive Technology* each having 6 articles produced in the year 2022 (Table 6). It can be noted that the production in *ACM International Conference Proceeding Series* did not start till the year 2010 whereas *Disability and Rehabilitation: Assistive Technology* had the production beginning from 2007.

3.3 Authors

Upon conducting an analysis at the Authors level, it was found that particular authors were more significant than others while also uncovering several bibliometric indicators such as country level collaborations. Lastly discussed was their production over a period of time (Fig. 3). In one document there were 34 authors.

In terms of volume of Production, there were 2 authors who showed a continuous trajectory over the years – these authors were Wolbring G and Zhu Z. In terms of number of articles produced (Fig. 4), Wolbring G led the publication with 6 articles

Table 2. Main Information. Source: own compilation

Description	Results
MAIN INFORMATION ABOUT DATA	
Timespan	1995:2022
Sources (Journals, Books, etc.)	170
Documents	225
Annual Growth Rate %	15.42
Document Average Age	5.42
Average citations per doc	19.3
References	1
DOCUMENT CONTENTS	
Keywords Plus (KP)	1427
Author's Keywords (KE)	747
AUTHORS	
Authors	803
Authors of single-authored docs	33
AUTHORS COLLABORATION	
Single-authored docs	34
Co-Authors per Doc	3.96
International co-authorships %	26.67
DOCUMENT TYPES	
article	97
article book	1
article review	1
book	18
book chapter	14
conference paper	67
note	1
review	26

whereas authors Annicchiarico R, Cortés U & Tiberio L had 5 articles each. The author collaboration network clusters have been made in accordance with the most productive authors and have been numbered from 1 to 10 based on the most productive authors as mentioned in the above figure.

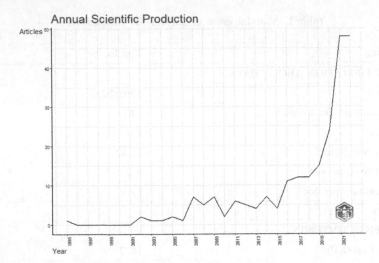

Fig. 1. Annual Scientific Production

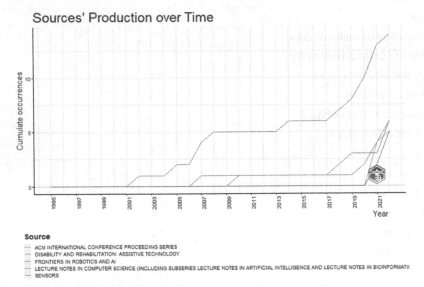

Fig. 2. Sources Scientific Production over time

Country, Production and Citations

In order to analyse the most productive countries through the country affiliation of authors, the country collaboration social networks, national and international collaboration, as well as productivity and citation networks were studied.

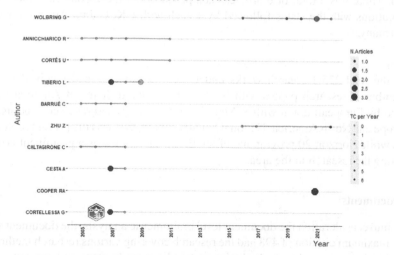

Fig. 3. Production of Top Authors over time

Table 3. Country Total Citation, Avg. Article Citation and Production

Country	TC	Average Article Citations	Frequency
USA	902	37.60	152
United Kingdom	587	53.40	52
Italy	564	47.00	45
Canada	148	13.50	47
Korea	104	26.00	35
France	61	15.20	11
Hungary	46	46.00	25
India	40	10.00	18
Germany	31	3.90	26
Belgium	25	25.00	27

It is worth noting that United States of America led the production of articles and the total citation count followed by the United Kingdom with 902 and 587 articles respectively (Table 3). Italy was a close third. One of the notable findings of the country citation was the emergence of a developing nation in contribution to the production of articles. India ranked 8th among the top 10 countries and had an increased number of articles than France but lacked in the total citation count and average article citations. The thickness of the connecting line explained the countries with the greatest number of collaborations. According to country collaborations, the most frequent collaborations were among USA and China. Italy and Spain also had equal number of collaborations

(Fig. 7). There was a total of 6 articles each produced by these countries. India had 2 collaborations with the states followed by 1 each with UK, China, Australia, Canada and Germany.

Funding

Out of the total 225 Research works under consideration, 87 had received funding. The number of research papers with funding from United States of America was 24 followed by European union with 8. Majority of the papers had received funding from the Europe and North America. National Science Foundation was the prominent funding agency while Horizon 20 programme of the European union was a prominent program supporting the research in the area.

3.4 Documents

Taking into consideration the document level bibliometric analysis, the document which had the maximum citation of 498 had the research covering various research methodologies and techniques used to study the interaction between humans and computers. The research provided practical insights and strategies for designing and conducting effective HCI studies, including user studies, usability testing, and data analysis. Other documents also focused on the field of robotics and its intersection with artificial intelligence. It covered the topics such as robot perception, planning, control, and learning algorithms. The article discusses advancements in intelligent robot systems and their applications in various domains, including manufacturing, healthcare, and autonomous vehicles. The document and the total citation count and total citation per year have been explained in the table below (Table 4).

Keywords Plus (KP) and Keywords (KE)

With consideration to the most frequently used Keyword Plus (KP) and Author Keywords (KE), minute differences can be seen in both of them. Majority of the author keywords except technology were identified with a corresponding ID in the database (Table 7 and Table 8) with minute differences in the occurrence. In order to understand the growth of Keyword Plus over time, a keyword plus graph was created (Fig. 4). It can be noted that from the year 2015 onwards, there has been a positive increase in research production with the keywords "Disabled Person", "Disabled Persons" and "Person with disability". To understand the growth of author keywords, a keyword graph was developed (Fig. 5) which showed that there has been an upward positive growth in the number of research production with the author keyword "Accessibility" which did not identify with a corresponding ID in the database.

Table 4. Most relevant - Total Citation (TC), Total Citation per year (TCY) and Source

Paper	Year	TC	TCY	Source
Lazar j, 2017, res methods in hum-comput interact	2017	498	71.14	Research methods in human-computer interaction
Burton mj, 2021, lancet global health	2021	320	106.67	The lancet global health
Kulyukin v, 2004, IEEE RSJ int conf intell robots and syst iros	2004	276	13.8	2004 IEEE/RSJ international conference on intelligent robots and systems (IROS)
Federici s, 2017, disabil rehabil	2017	199	28.43	Disability and rehabilitation
Emiliani pl, 2005, ibm syst j	2005	182	9.58	IBM systems journal
Gillespie a, 2012, j int neuropsychol soc	2012	181	15.08	Journal of the international neuropsychological society
Whyte em, 2015, j autism dev disord	2015	138	15.33	Journal of autism and developmental disorders
Ellis k, 2011, disabil and new media	2011	132	10.15	Disability and new media
Rizzo as, 2017, neuropsychology	2017	131	18.71	Neuropsychology
Higginbotham dj, 2007, aac augmentative altern commun	2007	98	5.76	AAC: augmentative and alternative communication

Thematic Analysis

Taking into account the total period from 1995 to 2022 and as mentioned in the methodology section, a strategic diagram explaining the underlying themes of the research have been developed. The size of the circle explains the prominence of the theme. Greater the size of the circle, greater is the prominence of the theme from among the research produced in the said period. With regards to the thematic analysis, the themes and the frequency of articles mentioned these themes have been explained in the table below (Table 5).

The thematic analysis was performed using a bi-dimensional thematic map to uncover the conceptual structure. This further explained the underlying basic themes, motor themes, niche theme and emerging themes (Kharo and Jain 2022) (Fig. 7). The basic themes generated from the thematic analysis was of "Person with disability" and had a cluster frequency of 82. The theme explores the application of AI-based assistive technologies for persons with disabilities. It discusses various AI techniques, such as machine learning and computer vision, utilized in areas such as communication, mobility, and independent living. The paper's in this theme highlights the benefits, challenges, and future directions of AI-based assistive tech for persons with disabilities (Sarker 2021).

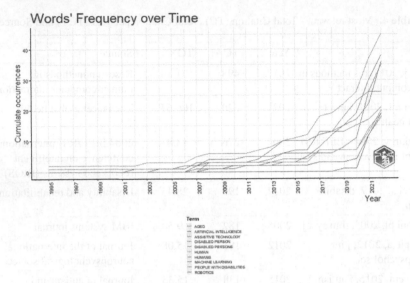

Fig. 4. Keywords Plus Growth

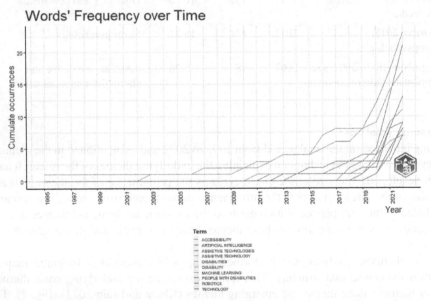

Fig. 5. Author Keywords Growth

The motor theme that emerged was of "Assistive Technology" which has a cluster frequency of 313. The theme focuses on machine learning approaches for predicting user preferences in assistive technologies for persons with disabilities. It discusses the use of various machine learning algorithms to personalize assistive systems based on user preferences and needs (Merabet et al.). It highlights the importance of considering individual user characteristics in designing AI-based assistive tech.

Author Keywords

Another important prominent motor themes that emerged were "Human" along with "Male" with a cluster frequency of 388 and 166 respectively as per the table below. The theme explores or rather focuses on improving human-computer interaction in AI-based assistive technologies for persons with disabilities. It discusses the importance of designing user-friendly interfaces, considering human factors, and incorporating user feedback in the development of assistive tech (Csapo et al.). The theme speaks about the significance of a human-centered approach to ensure the effectiveness and usability of AI-based solutions and a majority of the research has been done using males as the subjects of research.

The declining themes that emerged after conducting thematic analysis were "blind people", "visually impaired people" with a cluster frequency of 2 and 13 respectively. Niche themes "field trail", "computational linguistics" had a cluster frequency of 4 and 14 respectively (Fig. 6).

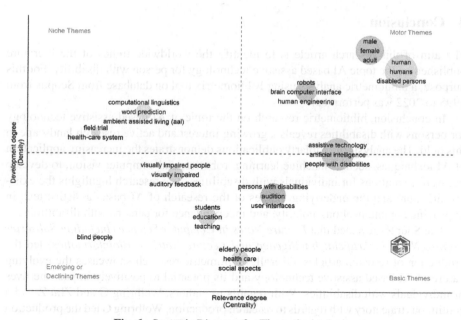

Fig. 6. Strategic Diagram for Thematic Analysis

Table 5. Cluster and Cluster Frequency

Cluster	Cluster Frequency
persons with disabilities	82
assistive technology	313
human	388
male	166
students	20
robots	38
elderly people	14
visually impaired people	13
computational linguistics	14
field trial	4
blind people	2

4 Conclusion

The aim of this research article is to identify the worldwide trends of the literature published on the topic AI based assistive technology for person with disability. For this purpose, a bibliometric analysis using Bibliometrix tool on database from Scopus from 1995 to 2022 was performed.

In conclusion, bibliometric research on the topic of AI-based assistive technology for persons with disabilities reveals a growing interest and active research landscape in this field. The analysis of scholarly publications demonstrates the increasing application of AI techniques, such as machine learning, robotics, and computer vision, to develop innovative solutions for individuals with disabilities. The research highlights the existing literature and the underlying themes of the research of AI-based assistive tech in improving communication, mobility, and independence for persons with disabilities.

The Sources revealed that *Lecture Notes in Computer Science (Including Subseries Lecture Notes in Artificial Intelligence and Lecture Notes in Bioinformatics)* led the production of research articles. Overall, bibliometric research showcases the evolving nature of AI-based assistive technology and its potential to positively impact the lives of individuals with disabilities. With regards to authors, Wolbring G and Zhu Z had a continuous trajectory with regards to research production. Wolbring G led the production frequency followed by Annicchiarico R, Cortés U & Tiberio L.

Interesting interactions were displayed in country collaborations and countries with greatest production. The Citation and the citation per year revealed interesting information about the countries conducting research on the said topic. USA ranked first in terms of productions but when it came to average article citation, It was United Kingdom which was ranked first followed by Italy and USA was ranked third. From among the developing nations, India did feature in the top countries producing literature on the said topic.

In relation to the author keywords and keywords plus, it can be noted there was little difference between keywords plus and author keywords. With regards to the documents, the most globally cited documents were Lazar J, 2017, Res Methods in Human-Computer Interaction with a total citation count of 498 followed by Burton Mj, 2021, Lancet Global Health with 320 total citations. The basic themes generated from the thematic analysis was of "Person with disability" and had a cluster frequency of 82. The motor theme that emerged was of "Assistive Technology" which has a cluster frequency of 313. The declining themes that emerged after conducting thematic analysis were "blind people", "visually impaired people" with a cluster frequency of 2 and 13 respectively.

4.1 Limitations and Future Research Directions

The study was conducted only on the literature published and indexed in Scopus database, there is a constant upgradation of the same and since the year 2023 was not considered as research period, the themes that are generated using bibliometric analysis especially the emerging ones may be subject to certain variations. The results may also vary if other globally accepted databases were also considered, and the literature published in them used for the analysis. However, the research results could help researchers delve deeper into the specific areas of the emerging themes and can be helpful in investigation of upcoming areas with the help of a specific focused group. The country collaboration and author collaboration analysis can also be helpful in policy framing or investigation into the capabilities of the collaborative countries and may come up with a strategic collaboration into one of the emerging themes of the study. The study can further include a time series analysis on the evolution of research using multiple globally accepted databases. This research offers the researchers the opportunity to further expand the databases used, the depth of analysis and even focus on a specific theme with AI and assistive technology especially pertaining to person with disability. The significance of AI-assistive technology development lies in their deployment, however research on how effectively this intervention is being used is limited. The availability, accessibility, and affordability dimensions of the AI assistive technologies are under addressed in the existing literature. While current body of literature touches on accessibility, aspects of commercialization, including affordability and availability, may be taken up for further research.

Further, the keyword analysis falls short of explaining the unique needs of the PWD being addressed through AI assistive technology interventions either on account of the lack of research in the area or use of appropriate keywords. A Systematic Literature Review (SLR) of the papers would help identify the same and strongly suggested.

Appendix A:

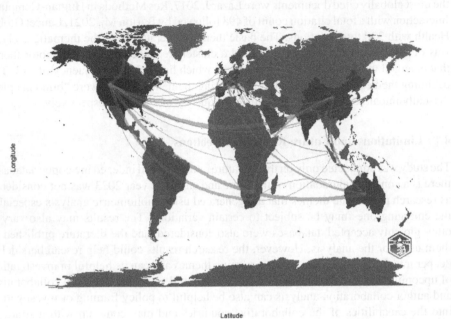

Fig. 7. Country Collaboration Network

Table 6. Most Relevant Sources

Sources	Articles
Lecture notes in computer science (including subseries lecture notes in artificial intelligence and lecture notes in bioinformatics)	14
ACM international conference proceeding series	7
Disability and rehabilitation: assistive technology	7
Sensors	6
Assistive technology	5
Frontiers in robotics and ai	5
Conference on human factors in computing systems - proceedings	4
Sustainability (Switzerland)	4
ACM transactions on accessible computing	3
AI and society	3

Table 7. Most Relevant Keyword Plus

Words	Occurrences
Human	45
Humans	38
Assistive technology	36
Artificial intelligence	35
Robotics	26
Aged	21
Disabled persons	20
Disabled person	19
People with disabilities	19
Machine learning	18

Table 8. Most Relevant Author Keywords

Words	Occurrences
Assistive technology	23
Accessibility	21
Disability	17
Artificial intelligence	13
Technology	11
Machine learning	9
Robotics	9
People with disabilities	8
Assistive technologies	7
Disabilities	7

References

1. Alper, S., Raharinirina, S.: Assistive technology for individuals with disabilities: a review and synthesis of the literature. J. Spec. Educ. Technol. **21**(2), 47–64 (2006)
2. Aria, M., Cuccurullo, C.: Bibliometrix: an R-tool for comprehensive science mapping analysis. J. Inform. **11**(4), 959–975 (2017)
3. Bartram, T., Cavanagh, J.: Re-thinking vocational education and training: Creating opportunities for workers with disability in open employment. J. Vocat. Educ. Train. **71**(3), 339–349 (2019)
4. Barua, P.D., et al.: Artificial intelligence enabled personalised assistive tools to enhance education of children with neurodevelopmental disorders—a review. Int. J. Environ. Res. Public Health **19**(3), 1192 (2022)

5. Broadbent, E., Stafford, R., MacDonald, B.: Acceptance of healthcare robots for the older population: review and future directions. Int. J. Soc. Robot. **1**, 319–330 (2009)

6. Callon, M., Courtial, J.P., Laville, F.: Co-word analysis as a tool for describing the network of interactions between basic and technological research: The case of polymer chemistry. Scientometrics **22**, 155–205 (1991)

7. Cobo, M.J., López-Herrera, A.G., Herrera-Viedma, E., Herrera, F.: Science mapping software tools: review, analysis, and cooperative study among tools. J. Am. Soc. Inform. Sci. Technol. **62**(7), 1382–1402 (2011)

8. Cortés, U., Annicchiarico, R., Vázquez-Salceda, J., Caltagirone, C.: e-Tools: the use of assistive technologies to enhance disabled and senior citizens' autonomy. In: EU-LAT Workshop on e-Health, Cuernavaca, Mexico, pp. 117–133 (2003)

9. Csapó, Á., Wersényi, G., Nagy, H., Stockman, T.: A survey of assistive technologies and applications for blind users on mobile platforms: a review and foundation for research. J. Multimodal User Interfaces **9**, 275–286 (2015)

10. Czaja, S.J., Ceruso, M.: The promise of artificial intelligence in supporting an aging population. J. Cogn. Eng. Decis. Mak. **16**(4), 182–193 (2022)

11. Gauthier, É.: Bibliometric analysis of scientific and technological research: a user's guide to the methodology (1998)

12. Glänzel, W., Debackere, K., Thijs, B., Schubert, A.: A concise review on the role of author self-citations in information science, bibliometrics and science policy. Scientometrics **67**(2), 263–277 (2006)

13. Howard, J., Fisher, Z., Kemp, A.H., Lindsay, S., Tasker, L.H., Tree, J.J.: Exploring the barriers to using assistive technology for individuals with chronic conditions: a meta-synthesis review. Disabil. Rehabil. Assist. Technol. **17**(4), 390–408 (2022)

14. Kamada, T., Kawai, S.: An algorithm for drawing general undirected graphs. Inf. Process. Lett. **31**(1), 7–15 (1989)

15. Kenigsberg, P.A., et al.: Assistive technologies to address capabilities of people with dementia: from research to practice. Dementia **18**(4), 1568–1595 (2019)

16. Khalilullah, K.I., Ota, S., Yasuda, T., Jindai, M.: Development of robot navigation method based on single camera vision using deep learning. In: 2017 56th Annual Conference of the Society of Instrument and Control Engineers of Japan (SICE), pp. 939–942. IEEE (2017)

17. Khare, A., Jain, R.: Mapping the conceptual and intellectual structure of the consumer vulnerability field: a bibliometric analysis. J. Bus. Res. **150**, 567–584 (2022)

18. Lee, C., Coughlin, J.F.: Perspective: older adults' adoption of technology: an integrated approach to identifying determinants and barriers. J. Prod. Innov. Manag. **32**(5), 747–759 (2015)

19. Lee, S., Yu, R., Xie, J., Billah, S.M., Carroll, J.M.: Opportunities for human-AI collaboration in remote sighted assistance. In: 27th International Conference on Intelligent User Interfaces, pp. 63–78 (2022)

20. Loi, M., Castriotta, M., Di Guardo, M.C.: The theoretical foundations of entrepreneurship education: how co-citations are shaping the field. Int. Small Bus. J. **34**(7), 948–971 (2016)

21. López-Pernas, S., Saqr, M., Apiola, M.: Scientometrics: a concise introduction and a detailed methodology for mapping the scientific field of computing education research. Past, Present Future Comput. Educ. Res.: Glob. Perspect., 79–99 (2023)

22. Malhotra, Y.: Why knowledge management systems fail: enablers and constraints of knowledge management in human enterprises. Handbook Knowl. Manage.1: Knowl. Matters, 577–599 (2004)

23. Mankoff, J., Hayes, G.R., Kasnitz, D.: Disability studies as a source of critical inquiry for the field of assistive technology. In: Proceedings of the 12th International ACM SIGACCESS Conference on Computers and Accessibility, pp. 3–10 (2010)

24. Medina, M.J., Lalanne, D., Baudet, C.: Human-computer interaction in artificial intelligence for blind and vision impairment: an interpretative literature review based on bibliometrics. In: IHM 2022: Proceedings of the 33rd Conference on l'Interaction Humain-Machine: Adjunct (IHM 2022 Adjunct) (2022)
25. Merabet, G.H., et al.: Intelligent building control systems for thermal comfort and energy-efficiency: a systematic review of artificial intelligence-assisted techniques. Renew. Sustain. Energy Rev. **144**, 110969 (2021)
26. Otte, E., Rousseau, R.: Social network analysis: a powerful strategy, also for the information sciences. J. Inf. Sci. **28**(6), 441–453 (2002)
27. Pons, P., Latapy, M.: Computing communities in large networks using random walks. J. Graph Algorithms Appl. **10**(2), 191–218 (2006)
28. Sarker, I.H.: Machine learning: algorithms, real-world applications and research directions. SN Comput. Sci. **2**(3), 160 (2021)
29. Scherer, M.J., Craddock, G., Mackeogh, T.: The relationship of personal factors and subjective well-being to the use of assistive technology devices. Disabil. Rehabil. **33**(10), 811–817 (2011)
30. Wehmeyer, M.L.: The Oxford Handbook of Positive Psychology and Disability. Oxford University Press, Oxford (2013)

Leaf Disease Detection Using Transfer Learning

Mohit Saharan and Ghanapriya Singh[✉]

Department of Electronics and Communication Engineering, National Institute of Technology, Kurukshetra 136119, Haryana, India
ghanapriya@nitkkr.ac.in

Abstract. The early diagnosis of leaf diseases is essential for maintaining the health and yield of crops. With advancements in deep learning and computer vision, transfer learning has emerged as a powerful technique for solving complex image classification problems. This paper presents a comparative analysis of three widely used convolutional neural network (CNN) models, namely ResNet, MobileNet, and VGG16, for leaf disease detection using transfer learning. The experimental results are evaluated according to a number of performance indicators, including as accuracy, precision, recall, and F1-score. Results show a test accuracy of 89.75%, 88.05% and 92.73%, respectively. The comparative analysis of the models provides insights into their respective strengths and weaknesses. Furthermore, visualizations of the confusion matrix and sample predictions offer a comprehensive understanding of their classification abilities. The paper also examines the practical implications of the models for real-time leaf disease detection. Factors such as inference time and computational resource requirements are considered to assess their suitability for deployment in real-world scenarios. The analysis tries to direct practitioners in choosing the best model for their particular application, taking the trade-off between accuracy and efficiency into consideration. In conclusion, this study provides a detailed comparative analysis of ResNet, MobileNet, and VGG16 models for leaf disease detection using transfer learning. The findings shed light on the performance, efficiency, and practical implications of these models, facilitating informed decision-making for researchers and practitioners working in the field of agricultural plant disease detection. The results also suggest potential avenues for future research and improvement in this domain.

Keywords: Transfer Learning · Leaf Disease Detection · VGG16 · MobileNet · ResNet

1 Introduction

Worldwide agricultural production and food security are seriously threatened by leaf diseases. Timely detection and accurate diagnosis of these diseases are essential for implementing effective control measures and minimizing crop losses.

S. Tiwari et al. (Eds.): AI4S 2023, CCIS 1907, pp. 44–58, 2023.
https://doi.org/10.1007/978-3-031-47997-7_4

Traditionally, manual inspection by experts has been employed to identify and classify leaf diseases. However, this approach is time-consuming, subjective, and often requires specialized knowledge.

In recent years, the field of computer vision and deep learning has witnessed remarkable advancements, revolutionizing various image classification tasks. Previous to deep learning algorithms, various machine learning algorithms like SVM [1], spatio-temporal features based algorithms [2], wavelet transforms [3], etc. and clustering algorithms like Local gravitational clustering [4] and frequency transformations [5] have been utilized in identification and classification problems. Deep learning [6] and ensemble deep learning algorithms [7] have paved a way for more robust detection and recognition. Convolutional neural networks (CNNs), a class of deep learning models, have demonstrated exceptional capabilities in visual recognition tasks, including object detection, image classification, and segmentation. Transfer learning, a technique that leverages pretrained models on large-scale datasets, has emerged as a powerful tool for addressing the challenges associated with limited labeled data and resource-intensive training processes.

Transfer learning involves utilizing the knowledge learned from one task and applying it to a different but related task. By leveraging the pre-trained weights of CNN models, which have been trained on massive datasets, the models can effectively learn and generalize from a smaller dataset specific to the task at hand. This approach not only reduces the need for extensive training data but also enables faster convergence and improved overall performance.

In the context of leaf disease detection, transfer learning has shown great potential in achieving accurate and efficient classification. By employing pretrained models and fine-tuning them on leaf disease datasets, the models can learn to distinguish between healthy and diseased leaves, as well as identify specific disease types. However, the selection of an appropriate CNN model for transfer learning is crucial, as different models have varying architectural characteristics, computational requirements, and performance trade-offs.

In this paper, we present a comparative analysis of three widely used CNN models, namely ResNet, MobileNet, and VGG16, for leaf disease detection using transfer learning. These models are chosen due to their popularity, proven performance in various computer vision tasks, and availability of pre-trained weights. The study aims to evaluate and compare the accuracy, efficiency, and suitability of these models for leaf disease detection.

In the following segments, the paper is divided into following sections. Related Work is explored in Sect. 2. Section 3 analyzes the model architecture viz. ResNet, MobilNet, VGG16. Section 4 examine the process of data preprocessing, model design consideration and complexity and training. Section 5 presents the results and discussion of all the deep learning models. Section 6 concludes the paper.

2 Related Work

The detection and classification of leaf diseases using computer vision techniques have gained significant attention in recent years. Researchers have explored various approaches, including traditional machine learning algorithms and deep learning methods, to address this important agricultural problem. In particular, transfer learning has emerged as a promising technique for improving the accuracy and efficiency of leaf disease detection models. In this section, we provide an overview of the related work in the field of leaf disease detection and transfer learning.

Early studies in leaf disease detection primarily focused on handcrafted feature extraction and traditional machine learning algorithms. These approaches often relied on extracting texture-based features, such as colour histograms, texture descriptors, and shape-based features, followed by classification using classifiers like Support Vector Machines (SVMs) or Random Forests. One of the model in [1] uses SVM to detect tomato diseases which shows an accuracy of 95.71%. While these methods achieved reasonable results, they heavily relied on manual feature engineering, which can be time-consuming and may not capture complex patterns in leaf images.

The advent of deep learning, particularly CNNs, has revolutionized the field of computer vision, including leaf disease detection. Without the requirement for manual feature engineering, deep learning models may automatically learn hierarchical features from raw image data. Authors in [8] uses CNN model to detect leaf diseases which shoes an accuracy of 86%. This has led to improved accuracy and robustness in detecting leaf diseases. However, deep learning models often require a large amount of labeled training data, which can be challenging to obtain for specific plant species and disease types.

Transfer learning has emerged as a valuable technique to address the limitations of limited training data in leaf disease detection. Transfer learning enables the models to initialize with learnt features and then fine-tune them on the target leaf disease dataset by utilizing pre-trained models, which are often trained on massive picture datasets like ImageNet. This approach enables the models to generalize better and achieve higher accuracy even with smaller training datasets.

Several studies have applied transfer learning in leaf disease detection and achieved promising results. For instance, authors in [9] utilized the VGG16 model with transfer learning for the identification of tomato leaf diseases which gives an accuracy of 88.6%. Their results showed that they outperformed the conventional machine learning approach. Similarly, Satwinder Kaur et al. (2018) employed transfer learning with the GoogLeNet model for detecting plant diseases from leaf images which give an acuuracy of 97.8% shown in [10]. The study reported high accuracy and highlighted the effectiveness of transfer learning in limited data scenarios. Deep learning based Inception ResNet is used to detect yellow rust wheat disease in [11].

A technique was developed by [12] to automatically detect plant disease in images of maize plants that were captured in the field. In order to create an autonomous corn detector, authors in [13] trained a deep convolution neural network using 1632 images of corn kernels. A method for identifying rice infections was put forth by Lu et al. [14] and was based on deep convolutional neural network (CNN) technology. Using a deep learning technique, Zhang et al. created a network for recognising images of farm equipment [15].For better understanding of CNN and Deep learning are in [16–19]. Some references for future work are in [20,21].

While transfer learning has shown great potential, the choice of the pre-trained model plays a crucial role in achieving optimal performance. Different pre-trained models, such as ResNet, MobileNet, and VGG16, have varying architectural complexities and computational requirements. However, a detailed comparative analysis of these models for leaf disease detection using transfer learning is limited in the existing literature. In summary, prior research has demonstrated the effectiveness of deep learning and transfer learning in leaf disease detection. Transfer learning, in particular, has emerged as a valuable technique for overcoming data limitations and improving classification accuracy. However, a comprehensive comparative analysis of different CNN models, including ResNet, MobileNet, and VGG16, for leaf disease detection using transfer learning is essential to identify the most suitable model for this specific task. This study aims to address this gap and provide insights into the performance and practical implications of these models in leaf disease detection.

3   Model Architecture and Design

In this section, we provide a detailed overview of the architectural characteristics and design considerations for the ResNet, MobileNet, and VGG16 models used in leaf disease detection. Understanding the architectural differences between these models is crucial for selecting the most suitable model for transfer learning and optimizing performance.

3.1   ResNet

ResNet, short for Residual Network, introduced a breakthrough concept of residual learning that addressed the degradation problem faced by deep neural networks. Utilizing residual blocks, which have skip connections or shortcuts that enable the network to learn residual functions, is the main principle of ResNet. These short cuts make it possible for the network to quickly learn how the input and desired output differ, which makes it easier to train deeper networks. ResNet architectures typically consist of multiple residual blocks, where each block consists of several convolutional layers, batch normalization, and rectified linear activation functions. The skip connections in ResNet enable the direct flow of information from one block to another, mitigating the vanishing gradient problem and allowing the network to learn more effective representations of leaf disease patterns.

3.2   MobileNet

MobileNet is designed specifically for mobile and embedded vision applications, aiming to strike a balance between model size and accuracy. It introduces depthwise separable convolutions, which decompose the standard convolution operation into separate depthwise and pointwise convolutions. Depthwise convolutions apply a single filter per input channel, followed by pointwise convolutions that combine the outputs from the depthwise convolutions. By using depthwise separable convolutions, MobileNet significantly reduces the number of parameters and computational complexity compared to traditional convolutional layers. This makes it well-suited for resource-constrained environments while still maintaining competitive accuracy. MobileNet architectures typically consist of several depthwise separable convolutional layers with optional pointwise convolutions and downsampling operations.

3.3   VGG16

The well-known deep CNN architecture VGG16 is renowned for both its efficiency and simplicity. VGG16 has 16 weight layers, including 13 convolutional layers and 3 fully linked layers. Small 3×3 filters with a stride of 1 are a characteristic of the convolutional layers, which aid in the preservation of spatial information. VGG16 architecture follows a sequential pattern with multiple convolutional layers, each followed by a max-pooling layer for downsampling. The final fully connected layers enable high-level feature extraction and classification. The design philosophy of VGG16 focuses on stacking multiple layers with small filters to increase the depth and capture more complex patterns in the input images.

3.4   Design Consideration

When selecting a model for leaf disease detection, several factors should be considered:

3.4.1   Model Complexity

The architectural complexity of the model affects the number of parameters and the computational requirements for training and inference. ResNet has a relatively higher complexity due to the presence of skip connections, while MobileNet and VGG16 have different design choices that influence their computational efficiency.

3.4.2   Accuracy and Efficiency

Different models trade-off accuracy and efficiency differently. ResNet, with its deep architecture, tends to achieve high accuracy but requires more computational resources. MobileNet, on the other hand, prioritizes efficiency and is suitable for resource-constrained environments. VGG16 strikes a balance between accuracy and efficiency, although it has a larger memory footprint compared to MobileNet.

3.4.3   Datasize

The size of the leaf disease dataset available for training is an important consideration. Transfer learning works best when the pre-trained model's knowledge can be effectively fine-tuned on the target dataset. If the training dataset is small, models with fewer parameters may generalize better and mitigate overfitting.

4   Dataset Preparation and Training

In this section, we describe the dataset used for leaf disease detection, which is sourced from Kaggle. The dataset contains images of diseases affecting potato, tomato, and pepper plants, specifically focusing on the detection of black diseases.

4.1   Kaggle Dataset: Potato, Tomato, and Pepper Black Diseases

The dataset used in this study is obtained from Kaggle [22] and comprises a collection of images depicting various black diseases affecting potato, tomato, and pepper plants. These diseases include but are not limited to early blight, late blight, black spot, black mold, and black rot. The dataset provides a diverse set of images that accurately represent the visual symptoms and patterns associated with black diseases in these plants. Each image is labeled with the specific disease type, allowing for supervised training and evaluation of the leaf disease detection models. Some of the images are (Fig. 1):

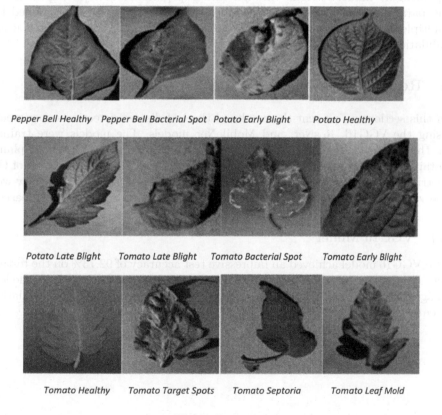

Pepper Bell Healthy   Pepper Bell Bacterial Spot   Potato Early Blight   Potato Healthy

Potato Late Blight   Tomato Late Blight   Tomato Bacterial Spot   Tomato Early Blight

Tomato Healthy   Tomato Target Spots   Tomato Septoria   Tomato Leaf Mold

Fig. 1. Dataset

4.2   Training the Models

Once the image dataset is extracted from Plant Village dataset from Kaggle using keras library, we can proceed with training the leaf disease detection models. The ResNet, MobileNet, and VGG16 models are particularly suitable for this task, as they have shown excellent performance in image classification tasks. Transfer learning is employed to leverage the pre-trained weights of these models. The initial layers, which have been trained on large-scale datasets like ImageNet, possess a strong ability to extract generic image features. Therefore, we freeze these initial layers and only train the final layers specific to the leaf disease detection task. By doing so, we can benefit from the pre-learned features and expedite the training process.

During training, optimization techniques Adam Optimizer with adaptive learning rate, weight decay, and momentum are commonly employed. Additionally, regularization methods like dropout and batch normalization can be applied to prevent overfitting and improve the model's generalization ability. The training process involves iteratively presenting the labeled images to the models, updating the model's weights through backpropagation, and adjusting the learning parameters to minimize the classification loss. The models are trained for multiple epochs until convergence, with the training progress monitored using validation metrics.

5   Results

In this section, we present the results of the leaf disease detection experiments using the VGG16, ResNet, and MobileNet models. The models were trained on the dataset comprising black diseases of potato, tomato, and pepper plants obtained from Kaggle. Accuracy, precision, recall, and F1 score are some of the metrics used to evaluate the models' performance. The outcomes show how well the various models work in correctly identifying and categorizing leaf diseases.

5.1   VGG16 Model

The VGG16 model achieved an impressive test accuracy of 92.73% on the testing set. This high accuracy indicates the model's ability to correctly classify the leaf images into their respective disease classes. The graph training and validation accuracy with epochs is as (Fig. 2):

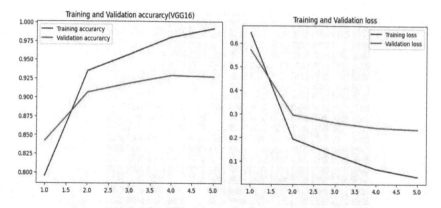

Fig. 2. Training and Validation accuracy vs Epochs(VGG16)

Additionally, the precision, recall, and F1 score for the VGG16 model were found to be consistently high, further confirming its strong performance. The VGG16 model's deep architecture and its ability to capture complex patterns in the images contribute to its superior performance. Confusion Matrix and classification report of the above model is as (Figs. 3 and 4):

	precision	recall	f1-score	support
Pepper__bell___Bacterial_spot	1.00	1.00	1.00	3
Pepper__bell___healthy	1.00	1.00	1.00	4
Potato___Early_blight	1.00	1.00	1.00	2
Potato___Late_blight	1.00	1.00	1.00	1
Tomato_Bacterial_spot	1.00	1.00	1.00	2
Tomato_Early_blight	1.00	1.00	1.00	1
Tomato_Late_blight	1.00	1.00	1.00	1
Tomato_Leaf_Mold	1.00	1.00	1.00	5
Tomato_Septoria_leaf_spot	1.00	1.00	1.00	4
Tomato_Spider_mites_Two_spotted_spider_mite	1.00	1.00	1.00	1
Tomato__Target_Spot	1.00	1.00	1.00	3
Tomato__Tomato_YellowLeaf__Curl_Virus	1.00	1.00	1.00	3
Tomato__Tomato_mosaic_virus	1.00	1.00	1.00	1
Tomato_healthy	1.00	1.00	1.00	1
accuracy			1.00	32
macro avg	1.00	1.00	1.00	32
weighted avg	1.00	1.00	1.00	32

Fig. 3. Classification Report of VGG16 model

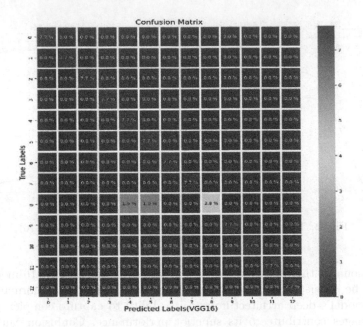

Fig. 4. Confusion Matrix of VGG16 model

5.2   ResNet Model

The ResNet model achieved an accuracy of 89.75% on the testing set, demonstrating its effectiveness in leaf disease detection. The graph training and validation accuracy with epochs is as (Fig. 5):

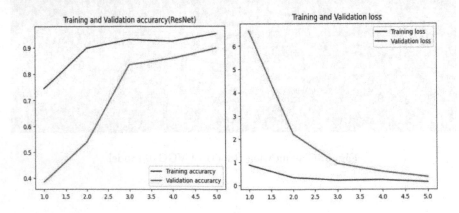

Fig. 5. Training and Validation Accuracy vs Epochs (ResNet)

While slightly lower than the accuracy of the VGG16 model, the ResNet model still exhibited strong performance. The precision, recall, and F1 score for the

ResNet model were also high, indicating its ability to accurately classify leaf diseases. Confusion matrix and classification report for the above model is as (Figs. 6 and 7):

	precision	recall	f1-score	support
Pepper__bell___Bacterial_spot	1.00	0.67	0.80	3
Pepper__bell___healthy	1.00	1.00	1.00	4
Potato___Early_blight	1.00	1.00	1.00	2
Potato___Late_blight	1.00	1.00	1.00	1
Potato___healthy	1.00	1.00	1.00	2
Tomato_Bacterial_spot	1.00	1.00	1.00	1
Tomato_Early_blight	1.00	1.00	1.00	1
Tomato_Late_blight	1.00	1.00	1.00	5
Tomato_Leaf_Mold	1.00	1.00	1.00	4
Tomato_Septoria_leaf_spot	1.00	1.00	1.00	1
Tomato_Spider_mites_Two_spotted_spider_mite	0.75	1.00	0.86	3
Tomato__Tomato_YellowLeaf__Curl_Virus	1.00	1.00	1.00	3
Tomato__Tomato_mosaic_virus	1.00	1.00	1.00	1
Tomato_healthy	1.00	1.00	1.00	1
accuracy			0.97	32
macro avg	0.98	0.98	0.98	32
weighted avg	0.98	0.97	0.97	32

Fig. 6. Classifiaction Report of ResNet Model

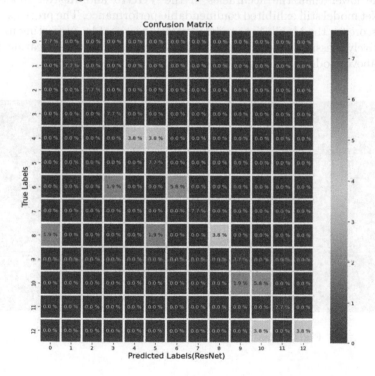

Fig. 7. Confusion Matrix of ResNet model

5.3   MobileNet Model

The MobileNet model achieved an accuracy of 88.05% on the testing set. The graph training and validation accuracy with epochs is as (Fig. 8):

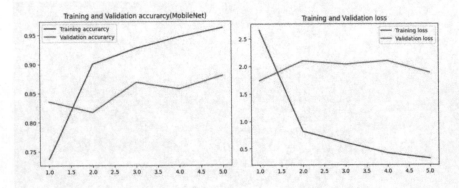

Fig. 8. Training and Validation Accuracy vs Epochs (MobileNet)

While lower than the accuracies of the VGG16 and ResNet models, the MobileNet model still exhibited commendable performance. The precision, recall, and F1 score for the MobileNet model were also reasonable, suggesting its ability to effectively classify leaf diseases. Confusion matrix and Classification Report for the above model is as (Figs. 9 and 10):

	precision	recall	f1-score	support
Pepper__bell__Bacterial_spot	1.00	0.33	0.50	3
Pepper__bell__healthy	1.00	0.75	0.86	4
Potato___Early_blight	1.00	0.50	0.67	2
Potato___Late_blight	0.50	1.00	0.67	1
Potato___healthy	0.00	0.00	0.00	0
Tomato_Bacterial_spot	0.67	1.00	0.80	2
Tomato_Early_blight	0.00	0.00	0.00	1
Tomato_Late_blight	1.00	1.00	1.00	1
Tomato_Leaf_Mold	1.00	0.60	0.75	5
Tomato_Septoria_leaf_spot	0.44	1.00	0.62	4
Tomato_Spider_mites_Two_spotted_spider_mite	0.33	1.00	0.50	1
Tomato__Target_Spot	1.00	0.67	0.80	3
Tomato__Tomato_YellowLeaf__Curl_Virus	1.00	0.33	0.50	3
Tomato__Tomato_mosaic_virus	1.00	1.00	1.00	1
Tomato_healthy	1.00	1.00	1.00	1
accuracy			0.69	32
macro avg	0.73	0.68	0.64	32
weighted avg	0.84	0.69	0.69	32

Fig. 9. Classification Report of MobileNet model

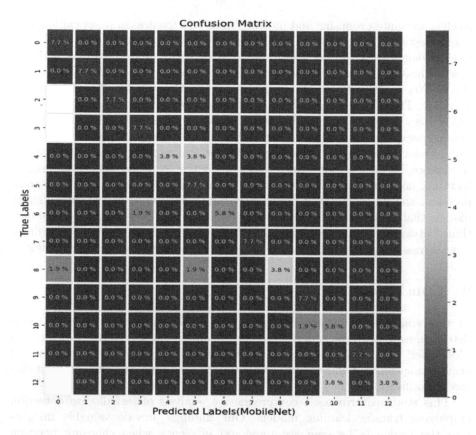

Fig. 10. Confusion Matrix of MobileNet model

The MobileNet model's focus on efficiency and its reduced computational complexity make it a suitable choice for resource-constrained environments, although it may sacrifice a small amount of accuracy compared to more complex models.

5.4   Performance Comparison

Comparing the results of the three models, the VGG16 model achieved the highest accuracy of 92.73%, followed by the ResNet model with 89.75% accuracy, and the MobileNet model with 88.05% accuracy. These results indicate that the VGG16 model performs slightly better in detecting and classifying leaf diseases in this particular dataset. However, it is important to consider the computational requirements and realtime deployment constraints when choosing the appropriate model. It is worth noting that achieving high accuracy rates in leaf disease detection is a challenging task due to the visual similarity of some disease symptoms and the potential presence of variations in lighting conditions, image quality, and plant species. Therefore, the accuracies achieved by all three models are quite promising and indicate their effectiveness in practical leaf disease detection applications. In conclusion, the VGG16, ResNet, and MobileNet models demonstrate strong performance in detecting and classifying leaf diseases. The VGG16 model achieved the highest accuracy of 92.73%, followed by the ResNet model with 89.75% accuracy and the MobileNet model with 88.05% accuracy. These results highlight the potential of deep learning models in accurately identifying and diagnosing leaf diseases, thus assisting in effective crop management and disease control strategies. The future of leaf disease detection using transfer learning is promising, with several exciting research directions to explore. By focusing on diverse datasets, active learning, weakly supervised learning, domain adaptation, few-shot learning, multi-modal data, model compression, and interpretability, researchers can develop more robust and efficient models that contribute to sustainable agriculture and food security. Collaborations between researchers, plant pathologists, and agricultural experts will be crucial in addressing the challenges and seizing the opportunities in this field.

6 Conclusion

In this study, we embarked on a comprehensive comparison of leaf disease detection using three prominent transfer learning models: VGG16, ResNet, and MobileNet. Our primary objective was to assess their performance in terms of accuracy, speed, and efficiency in combating the pressing challenge of plant disease identification.

This study contributes a comprehensive analysis of leaf disease detection employing transfer learning models. Our findings provide valuable insights into the trade-offs between accuracy and efficiency when choosing between VGG16, ResNet, and MobileNet. As the agricultural sector increasingly relies on technology-driven solutions, the results presented here offer guidance for researchers and practitioners seeking to implement effective disease detection systems for the betterment of crop yield and food security.

References

1. Sathyaa, S.P.A., Ramakrishnana, S., Shafreena, M.I., Harshini, R., Malinia, P.: Optimal plant leaf disease detection using SVM classifier with Fuzzy System. In: Workshop on Intelligent Systems, 22–24 April 2022, Chennai, India (2022)
2. Bijalwan, V., Semwal, V.B., Singh, G., Mandal, T.K.: HDL-PSR: modelling spatiotemporal features using hybrid deep learning approach for post-stroke rehabilitation. Neural Process. Lett. 1–20 (2022). https://doi.org/10.1007/s11063-022-10744-6
3. Singh, G., Singh, R.K., Saha, R., Agarwal, N.: IWT based iris recognition for image authentication. Procedia Comput. Sci. **171**, 1868–1876 (2020)
4. Kaloni, S., Singh, G., Tiwari, P.: Nonparametric damage detection and localization model of framed civil structure based on local gravitation clustering analysis. J. Build. Eng. **44**, 103339 (2021)

5. Kaloni, S., Tiwari, P., Singh, G.: User-defined high impulsive frequency acquisition model for mechanical damage identification. In: Proceedings of the Institution of Mechanical Engineers, Part K: Journal of Multi-body Dynamics, p. 14644193231157176 (2023)

6. Singh, G., Chowdhary, M., Kumar, A., Bahl, R.: A probabilistic framework for base level context awareness of a mobile or wearable device user. In: 2019 IEEE 8th Global Conference on Consumer Electronics (GCCE), pp. 217–218. IEEE (2019)

7. Bijalwan, V., Semwal, V.B., Singh, G., Crespo, R.G.: Heterogeneous computing model for post-injury walking pattern restoration and postural stability rehabilitation exercise recognition. Expert. Syst. **39**(6), 12706 (2022)

8. Shobana, M., Vaishnavi, S., SP, P.K., Madhumitha, K.P., Nitheesh, C., Kumaresan, N.: Plant disease detection using convolution neural network. In: 2022 International Conference on Computer Communication and Informatics (ICCCI), pp. 1–5 (2022). https://doi.org/10.1109/ICCCI54379.2022.9740975

9. Alok Kumar, A.K.: Plant disease detection using vgg16. Int. J. Creat. Res. Thoughts **11**(2), c770–c775 (2023)

10. Satwinder Kaur, R.V. Garima Joshi: Plant disease classification using deep learning google net model. Int. J. Innov. Technol. Explor. Eng. **8**(2), 319–322 (2019)

11. Zhang, X., et al.: A deep learning-based approach for automated yellow rust disease detection from high-resolution hyperspectral UAV images. Remote Sens. **11**, 1554 (2019). https://doi.org/10.3390/rs11131554

12. DeChant, C., et al.: Automated identification of northern leaf blight- infected maize plants from field imagery using deep learning. Phytopathology® **107**(11), 1426–1432 (2017). https://doi.org/10.1094/PHYTO-11-16-0417-R

13. Ni, C., Wang, D., Vinson, R., Holmes, M., Tao, Y.: Automatic inspection machine for maize kernels based on deep convolutional neural networks. Biosyst. Eng. **178**, 131–144 (2019). https://doi.org/10.1016/j.biosystemseng.2018.11.010

14. Lu, Y., Yi, S., Zeng, N., Liu, Y., Zhang, Y.: Identification of rice diseases using deep convolutional neural networks. Neurocomputing **267**, 378–384 (2017). https://doi.org/10.1016/j.neucom.2017.06.023

15. Zhang, Z., Liu, H., Meng, Z., Chen, J.: Deep learning-based automatic recognition network of agricultural machinery images. Comput. Electron. Agric. **166**, 104978 (2019). https://doi.org/10.1016/j.compag.2019.104978

16. Singh, G., Chowdhary, M., Kumar, A., Bahl, R.: A personalized classifier for human motion activities with semi-supervised learning. IEEE Trans. Consum. Electron. **66**(4), 346–355 (2020). https://doi.org/10.1109/TCE.2020.3036277

17. Tanwar, R., Chetia Phukan, O., Singh, G., Mishra Tiwari, S.: CNN-LSTM based stress recognition using wearables (2023)

18. Phukan, O.C., Singh, G., Tiwari, S., Butt, S.: An automated stress recognition for digital healthcare: towards e-governance. In: Ortiz-Rodríguez, F., Tiwari, S., Sicilia, M.-A., Nikiforova, A. (eds.) Electronic Governance with Emerging Technologies, pp. 117–125. Springer, Cham (2022). https://doi.org/10.1007/978-3-031-22950-3_10

19. Kanagaraj, N., Hicks, D., Goyal, A., Tiwari, S., Singh, G.: Deep learning using computer vision in self driving cars for lane and traffic sign detection. Int. J. Syst. Assur. Eng. Manage. **12**(6), 1011–1025 (2021). https://doi.org/10.1007/s13198-021-01127-6

20. Lu, Y., Chen, D., Olaniyi, E., Huang, Y.: Generative adversarial networks (GANs) for image augmentation in agriculture: a systematic review. Comput. Electron. Agric. **200**, 107208 (2022). https://doi.org/10.1016/j.compag.2022.107208

21. Agarwal, N., Sondhi, A., Chopra, K., Singh, G.: Transfer learning: survey and classification. In: Tiwari, S., Trivedi, M.C., Mishra, K.K., Misra, A.K., Kumar, K.K., Suryani, E. (eds.) Smart Innovations in Communication and Computational Sciences. AISC, vol. 1168, pp. 145–155. Springer, Singapore (2021). https://doi.org/10.1007/978-981-15-5345-5_13
22. Emmanuel, T.O.: PlantVillage Dataset. & figshare https://www.kaggle.com/datasets/emmarex/plantdisease (2018)

Automated Scene Recognition for Environmental Monitoring: A Cluster Analysis Approach using Intel Image Classification Dataset

Yoginii Waykole[✉], Yashasvi Kanathey[✉], Vaani Goel[✉], Anupkumar Bongale, Prachi Kadam, and Kalyani Kadam

Department of Artificial Intelligence and Machine Learning, Symbiosis Institute of Technology, Pune, Symbiosis International (Deemed University), Lavale, Pune 412115, Maharashtra, India
yoginiiw10@gmail.com, kanatheyashasvi.18@gmail.com,
vaanigoel10@gmail.com, {anupkumar.bongale,prachi.kadam,
kalyanik}@sitpune.edu.in

Abstract. Automated scene identification is a crucial component of environmental monitoring since it makes it possible to analyse and manage various landscapes effectively. In this study, we examine how different clustering methods perform when used to the Intel Image Classification dataset, which consists of six different classes: roads, buildings, glaciers, sea, mountains, and woods. Our main objective is to find significant patterns in the information that will help with proper scene categorization.

We used PCA (Principal Component Analysis) to improve clustering effectiveness by reducing the dimensionality of the picture information. Then, we used K-Means, Agglomerative, BIRCH, DBScan, and Spectral clustering, five well-known clustering techniques. To evaluate the effectiveness of each method, we employed three assessment metrics: silhouette score, Calinski-Harabasz index, and Davies-Bouldin index.

Despite only creating two clusters, our results showed that Spectral clustering consistently beat the other methods in all three assessment measures. This outcome highlights the effectiveness of spectral clustering in capturing non-linear features inherent in the data, resulting in a better comprehension of the underlying scene categories. K-Means, Agglomerative, and BIRCH also produced four clusters, but DBScan found a cohesive structure with just one cluster, showing the dataset had particular properties.

In order to handle a modest number of clusters, K-Means, Agglomerative, and BIRCH were used. DBScan was used to find density-based clusters and outliers. The potential of spectral clustering to reveal non- linear patterns is why it was chosen.

The development of automatic scene identification for environmental monitoring is aided by our study. We emphasise the significance of taking into account non-linear correlations in picture data. Our findings sets the door for more in-depth investigation of feature extraction methods to improve scene categorization precision, ultimately assisting wise and long-lasting environmental monitoring practises.

© The Author(s), under exclusive license to Springer Nature Switzerland AG 2023
S. Tiwari et al. (Eds.): AI4S 2023, CCIS 1907, pp. 59–73, 2023.
https://doi.org/10.1007/978-3-031-47997-7_5

In conclusion, this paper proposes a thorough cluster analysis strategy for automated scene identification, demonstrating the potency of Spectral clustering for locating non-linear features in the Intel Image Classification dataset. The research's findings and new knowledge can help advance automatic scene identification technology, which will help environmental monitoring and associated applications.

Keywords: Clustering · Image Recognition · Spectral · BIRCH · Agglomerative · DB Scan · K-Means

1 Introduction

Unsupervised machine learning fundamentally arranges data points into comparable clusters or subgroups based on their properties or qualities. This process is known as clustering. In this work, we used the Intel Image Classification dataset to apply a number of well-liked clustering techniques. We employed K-means, agglomerative hierarchical clustering, DBSCAN, BIRCH, and Spectral among other techniques. A common partitioning clustering approach is K-means, which seeks to reduce the sum of squared distances between data points and the centroids of each cluster. The Intel Image Classification dataset's photos were divided into a predetermined number of clusters using the K-means technique. Utilizing measurements like the silhouette score, which gauges the effectiveness of the clustering based on the compactness and separation of the clusters, we assessed the efficacy of the K-means method.

Data points that are similar to one another in terms of their feature space are grouped together by the density-based clustering method DBSCAN. The DBSCAN algorithm was used to group the photos based on their density, and metrics like the silhouette score and the quantity of noise points were used to assess the programme's success.

BIRCH is a clustering technique that creates a tree-like structure of clusters to handle huge datasets. We clustered the photos using the BIRCH algorithm according to their feature space, and we measured the method's effectiveness using metrics like the number of clusters and the average silhouette score.

A layered cluster structure is created using the hierarchical clustering method known as agglomerative hierarchical clustering. The pairwise distance matrix, which gauges how similar two photos are based on their attributes, was used to cluster the images using the agglomerative hierarchical clustering technique. We measured the agglomerative hierarchical clustering algorithm's performance using measures like the cophenetic correlation coefficient. Popular unsupervised machine learning techniques include spectral clustering, which entails reducing the dimensions of the data using graph theory in order to make the clusters easier to see. With this approach, a graph is created from the data points, and the graph is then divided into clusters using eigenvectors and eigenvalues. On datasets like the Intel picture Classification dataset, where the photos are of outside, natural settings, it has been shown that spectral clustering performs well for picture classification tasks. For purposes like environmental monitoring, it may be possible to use spectral clustering to automatically distinguish and categorise scenes in such datasets.

Understanding the dynamic connections between ecosystems and human activity depends critically on environmental monitoring. There is a rising need for cutting-edge technology that can effectively analyze and interpret massive volumes of environmental data as public awareness of environmental concerns increases. Automated Scene Recognition is one such technique that uses machine learning to recognize and cluster scenes from images in various environmental circumstances.

Efficient Scene Categorization is a crucial aspect of our research, made possible by cluster analysis. Leveraging visual similarities, cluster analysis automates the categorization of environmental images into coherent classes, including forests, urban landscapes, oceans, and more. This method effectively organizes the large amount of environmental data by grouping related images together, speeding up the data analysis procedure, and saving significant time and effort.

2 Literature Review

Deep learning methods are used on remote sensing data to recognize scenes for environmental monitoring, and address the usage of several architectures, including Convolutional Neural Networks (CNNs), in applications for environmental monitoring [18].

The authors investigate the performance of the K-Means in terms of its running time and clustering quality under various conditions, including different initialization methods and number of clusters [1].

The author suggests two significant upgrades to DBSCAN in his updated version. The technique first modifies the density threshold according on the size of the dataset, allowing for more precise clustering results on datasets with different densities. Second, the algorithm adds a dynamic threshold for the minimum number of points necessary to form a cluster, increasing clustering accuracy even more [2]. The BIRCH algorithm, which makes use of feature vectors and a hierarchical approach to address clustering issues. The clustering feature, CF-tree, and the two-phase clustering technique, among other algorithmic parts, are all explained in detail [3].

The agglomerative clustering algorithm and its modifications are examined. It offers perceptions on trade-offs between temporal complexity and cluster quality as well as the two. The authors investigate methods for striking a balance between these trade-offs, such as utilizing heuristics to speed up the process without significantly reducing quality [8].

Demonstration of how different clustering algorithms perform on diverse dataset types using a variety of measures, including the silhouette coefficient. It provides insights into the benefits and drawbacks of each algorithm as well as recommendations for choosing the best algorithm for particular applications [12].

A summary of clustering techniques for assembling related objects or data points. Numerous clustering techniques exist, including model-based, partition-based, hierarchical, and density-based ones. The data and the analysis's objectives influence the algorithm that is used. There are numerous uses for clustering, and it is crucial to evaluate the outcomes using metrics like the silhouette coefficient and purity [15].

Analysis of spectral clustering, including a demonstration that it can successfully group data points in the face of noise and a discussion of how it compares to other clustering techniques like K-means clustering and hierarchical clustering [16].

In this research, we present an innovative approach to Automated Scene Recognition for Environmental Monitoring. Leveraging the power of cluster analysis, our goal is to accurately classify scenes using the widely recognized Intel Image Classification Dataset. Our research contributes to the advancement of automated scene recognition technologies, offering an alternative approach to deep learning techniques that are prevalent in the literature. In addition, comparative analysis of different clustering algorithms to accurately classify scenes using the Intel Image Classification Dataset.

3   Data Preprocessing and EDA

3.1   About the Dataset

A set of pictures called the Intel Image Classification Dataset is frequently used for developing and evaluating computer vision models for image classification applications. It has 25,000 photos, which are divided into six groups according to size and resolution: houses, forests, glaciers, mountains, sea, and streets.

The dataset was made freely accessible for use by academics and developers in their projects and was produced by Intel as part of their AI for Social Good effort. A group of human annotators carefully identified the photographs after they were gathered from multiple sources, including Flickr and Google Images, to verify correctness.

This dataset was taken from Kaggle (https://www.kaggle.com/datasets/puneet6060/intel-image-classification).

3.2   Data Transformation

The data preprocessing pipeline includes a crucial step called data transformation that helps machine learning models perform better. The following standard data transformation methods have been performed:

Resizing: The dataset's picture sizes may vary, which might make it challenging to train a machine-learning model. The input data may be standardized by resizing the photographs to a constant size, which can enhance model performance.

Normalization: The Intel Image Classification dataset's normalization method would result in a version of the dataset where the picture pixel values have been rescaled to a standard scale, often between 0 and 1 or −1 and 1. While normalizations serve to lessen the influence of variations in brightness and contrast between pictures, it is frequently done to enhance the performance and stability of machine learning algorithms.

Gray-scaling: By converting pictures from RGB to grayscale, one may decrease the number of dimensions in the input data and speed up the model-training procedure.

Feature extraction: By supplying more insightful input data, techniques like principal component analysis, Singular Vector Decomposition, and color histograms may be used to extract features from the photos and enhance model performance. These standard data transformation methods can be used to enhance the performance of machine learning models on the Intel Picture Classification dataset.

3.3 Dimensionality Reduction with PCA for Improved Clustering Efficiency

For high-dimensional datasets, such as images, dimensionality reduction is a crucial preprocessing step. To improve clustering performance and lower computing costs, it is essential to reduce the dimensionality of the Intel image dataset, which has a significant number of features.

We used Principal Component Analysis (PCA) on the Intel image dataset to enhance clustering efficiency and minimize computational cost of the techniques. A well-known method for dimensionality reduction in machine learning is PCA, which reduces the amount of information in high-dimensional data while retaining the most crucial details. On the PCA data, we clustered using various techniques.

4 Clustering Methods for Scene Recognition

4.1 K Means Clustering Technique

An effective unsupervised machine learning approach for clustering a given set of data points based on their similarities is called K-means clustering. By minimizing the sum of squared distances between the data points and their respective cluster centroids, the algorithm attempts to divide the data into K clusters, where K is a specified number. The K-means clustering algorithm's operation can be summed up as follows:

Initialization: Choose K initial centroids at random from the given dataset.

Assignment: Based on its Euclidean distance, each data point is given to the nearest centroid.

Update: Recalculate each cluster's centroids by averaging all the data points belonging to that cluster. Repeat until the centroids stop moving or a predetermined number of iterations has been reached.

Output: The final K clusters and their respective centroids are produced by the algorithm.

The sum of squared distances between the data points and their respective cluster centroids are minimized using the K-means algorithm through an iterative process. The algorithm determines the Euclidean distance between each data point and the centroids at each iteration and places the data point's coordinates on the nearest centroid. The algorithm updates the centroids by figuring out the mean of all the data points in the cluster once each data point has been assigned to its corresponding cluster. Until the centroids stop moving or the allotted number of iterations has been reached, the assignment and update procedures are repeated.

The algorithm may be sensitive to the starting centroids used and converge to a less-than-ideal result. To discover the ideal answer, numerous runs with various initializations are frequently carried out (Figs. 1 and 2).

The outcome of the ideal number of clusters was discovered to be four using the elbow approach. The resulting clustering revealed a noticeable division of the data points into 4 clusters.

K-means clustering was selected because of its versatility for unsupervised learning, efficacy with large datasets, and accurate picture categorization.

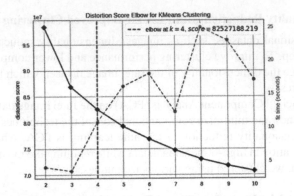

Fig. 1. Shows the ideal number of clusters for K Means Clustering

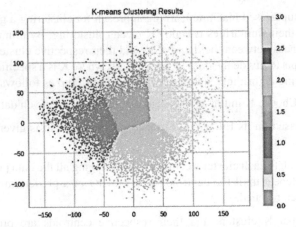

Fig. 2. Represents the K Means clustering of the environmental scenes

4.2 Agglomerative Clustering

Agglomerative clustering is a hierarchical clustering technique that iteratively clusters similar data points based on a distance metric into clusters until all the data points are combined into a single cluster or the appropriate number of clusters are found.

The technique begins by constructing N clusters, where N refers to the total number of data points and allocating each data point to its own cluster. Then, using a predetermined distance metric (such as Euclidean distance or cosine similarity), Each time, it calculates the separation between each pair of clusters. The total number of clusters is then decreased by one by joining the two clusters that are closest to one another. This procedure is continued until a stopping condition, such as the desired number of clusters or a threshold distance is met (Fig. 3).

The clustering graph shows 4 clusters were created as a result of the agglomerative clustering. The graph demonstrates the different clustering of the data points into 4 groups, indicating that there may be significant variances across the classes.

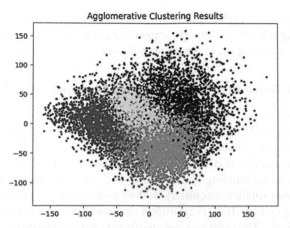

Fig. 3. Agglomerative clustering for 4 Clusters.

The dendrogram, a tree-like diagram that depicts the hierarchical relationships between the clusters at each iteration, can be used to visualize the agglomerative clustering technique. The interior nodes of the dendrogram represent fused clusters, whereas the leaves represent distinct data points.

The dendrogram shows that the agglomerative clustering on the Intel Image Classification dataset created four unique clusters. The photographs inside each cluster were found to exhibit comparable visual characteristics and portray the same outdoor landscapes when the images were plotted according to their cluster designations. For instance, one cluster had pictures of woods and other natural settings, while another cluster featured pictures of mountains and other geological features. These patterns imply that agglomerative clustering may efficiently classify photographs according to their visual characteristics and recognise comparable scenes in outside natural settings (Fig. 4).

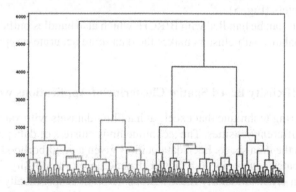

Fig. 4. Shows the dendogram of Agglomerative Clustering.

Agglomerative clustering is an effective and popular method for exploring and spotting patterns in large, complicated datasets in machine learning and data analysis.

Made the use of the Agglomerative clustering algorithm's hierarchical structure and effective handling of huge datasets. We sought to identify natural groups at a finer degree of granularity by deciding to cut the tree at 4 clusters.

4.3 BIRCH (Balanced Iterative Reducing and Clustering Using Hierarchies)

An effective clustering approach for huge datasets is BIRCH (Balanced Iterative Reducing and Clustering utilizing Hierarchies). It is a hierarchical clustering technique that organizes the data into subclusters in a tree-like form.

The dataset is iteratively divided into a tree of subclusters via the BIRCH method. The process begins by building a tree with a single node for the entire dataset. Then breaks the data into smaller subclusters, adding nodes to the tree recursively. The "CF Tree" (Clustering Feature Tree) is a clustering feature that the technique uses to conduct clustering effectively. The technique can swiftly calculate distances between subclusters thanks to the CF tree, a tree data structure that maintains a condensed summary of the information in each subcluster.

Create the CF Tree: The CF tree is created by scanning the dataset and adding new data points to the tree. A list of the data points in each leaf node of the tree, which represents a subcluster of the data, is provided. The number of data points, the subcluster's centroid (mean), and the sum of the squared distances of the data points from the centroid are all included in the summary.

Cluster the CF Tree: Starting at the root node, the CF tree is recursively clustered. The maximum distance between subclusters that can be merged is specified by a clustering threshold, which is used to do the clustering. The algorithm groups subclusters whose centroids are closer than the clustering threshold at each level of the tree.

Assign Data Points to Clusters: The procedure then uses a distance metric to allocate each data point to the closest subcluster after the CF tree has been clustered. This yields a final set of clusters (Fig. 5).

Large datasets can be handled with BIRCH, which also handles outliers well. Having the capacity to balance sub-clusters makes the data more accurately represented.

4.4 DBSCAN (Density Based Spatial Clustering of Applications with Noise)

A popular clustering technique that excels at handling datasets with various geometries and clusters of different densities. The technique finds clusters of data points that match to dense areas in the data space. Each data point is given a neighborhood by DB-SCAN, which then allocates each point to a cluster depending on the density of points in that neighborhood, in order to identify these regions. A point is specifically referred to as a core point if there are a minimum number of other points within a predetermined radius of it. After then, the cluster is created by incorporating all of the core points and any non-core points that are located nearby.

Finding patterns and structures in the data that might distinguish one class from another was the goal of the DB-SCAN clustering technique. Surprisingly, the clustering

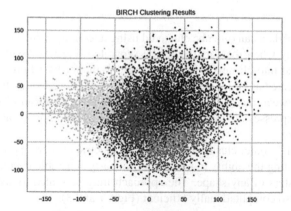

Fig. 5. Represents BIRCH Clustering.

technique only produced one cluster, showing that the data lacked any recognizable structures or patterns. This unexpected result raises the possibility that the dataset is homogeneous or that there are commonalities among all classifications.

The fact that DB-SCAN only produced one cluster indicates that this method may not be appropriate for this particular dataset or situation, and it may be necessary to investigate alternate clustering methods in order to gain a more thorough knowledge of the data (Fig. 6).

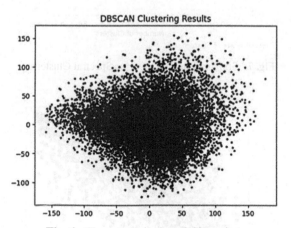

Fig. 6. Shows Density based Clustering.

DBScan has the ability to recognise clusters of various forms and is also capable of spotting outliers. Given that it only identified one cluster for our sample, it is possible that the distribution of our data deviates from the assumptions made by conventional clustering.

4.5 Spectral

High-dimensional data can often be clustered using a technique known as spectral clustering, particularly when the data cannot be separated linearly. A similarity graph of the data is divided into clusters as part of the graph theory-based procedure. The data points are represented as nodes in the graph, and the edges connecting them stand in for the similarity between the nodes. Any function, such as Euclidean distance or cosine similarity, that expresses the idea of proximity between data points can be used as the similarity measure.

In comparison to other clustering techniques, spectral clustering provides a number of benefits, including the ability to handle non-linearly separable data and the capacity to recognize clusters of any shape. The approach may be scaled to accommodate big datasets and is also computationally efficient (Figs. 7 and 8).

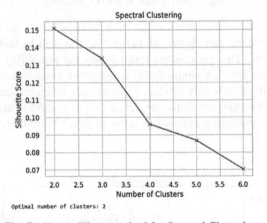

Optimal number of clusters: 2

Fig. 7. Shows Elbow method for Spectral Clustering.

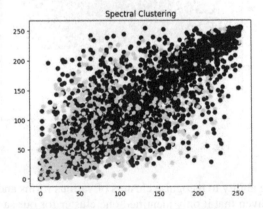

Fig. 8. Spectral Clustering of Data Points.

The fact that Spectral clustering outperformed other methods with just 2 clusters shows that this algorithm is better able to capture the underlying structure of the data.

It suggests that there are unique patterns among the data points in the Intel Image Classification dataset that are well represented by just two groups.

5 Evaluation of Clustering Algorithms

5.1 Silhouette Score

A popular evaluation metric for clustering algorithms is the silhouette score. It runs from -1 to 1 and measures how well each data point is segregated from other clusters. A point receives a score of 1, 0 indicates that it is on the boundary, and -1 indicates that it was incorrectly identified.

The following formula is used to determine the silhouette score:

$$s(i) = \frac{\max(a(i) \cdot b(i))}{b(i) - a(i)} \tag{1}$$

where a(i) is the average distance between point i and all other points in the same cluster, b(i) is the minimum average distance between point i and all other points in a different cluster, and s(i) is the silhouette score for the i-th data point.

The average of the silhouette scores of all the data points in all clusters is used to compute the overall silhouette score. Given that the data points are cleanly separated from one another and clearly belong to their respective clusters, a higher silhouette score denotes better grouping.

The silhouette score has the benefit of being fairly simple to interpret and comprehend. It can be used to evaluate the effectiveness of various clustering techniques and determine how many clusters are best for a certain dataset.

The silhouette score does have some restrictions, though. Although this may not always be the case with real-world data, it makes the assumption that clusters are convex and isotropic. Additionally, the score could not always serve as the optimum evaluation metric for all datasets due to its sensitivity to the clustering algorithm's initial configuration. As a result, it's crucial to combine the silhouette score with other evaluation metrics and carefully choose the best clustering algorithm for a particular dataset.

The Spectral clustering technique in this comparison has the highest silhouette score of 0.150, showing that it outperforms the others at categorizing data points into discrete clusters. Spectral clustering is a graph-based clustering method that groups the data points based on their connections in a lower-dimensional space after the data has been transformed into that space.

The second-highest silhouette score 0.118 belongs to K-means clustering, which is efficient in grouping data points into clusters. A straightforward and popular clustering algorithm called K Means divides the data into K clusters, where K is the user-specified number of clusters.

Despite having a lower silhouette score than K Means and spectral clustering 0.074, agglomerative clustering still shows acceptable performance. A hierarchical clustering algorithm called agglomerative clustering works by repeatedly joining the clusters or pairs of data points that are most similar to one another.

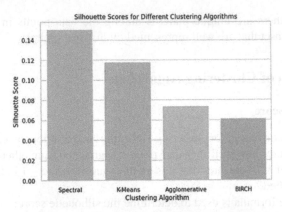

Fig. 9. Shows the comparative analysis of clustering algorithm on the basis of Silhouette Score.

The silhouette score of 0.066 for birch clustering is lower than that of the other techniques presented so far. Birch clustering is a clustering algorithm that divides the data into smaller clusters by creating a tree-like data structure (Fig. 9).

After comparing silhouette scores, it can be concluded that spectral and K Means clustering outperform the other algorithms at dividing data points into separate clusters. It's crucial to remember that the best clustering technique for a given dataset can change based on the information and the precise issue that needs to be resolved.

Therefore, when choosing the best method for a certain task, it's crucial to compare several clustering algorithms and take into account evaluation metrics other than the silhouette score.

5.2 Davis-Bouldin Score

A popular metric for assessing the calibre of clustering findings is the Davies- Bouldin score (DBI). It is the average of the highest similarity between each cluster and its most similar cluster, where similarity is measured by the Euclidean distance between the cluster centroids. It quantifies the ratio of the within-cluster distance to the between-cluster distance. Better clustering performance is indicated by a lower value of DBI since it suggests that the clusters are more compact and well-separated from one another.

The fact that the Davies-Bouldin index is suited for unsupervised learning tasks and does not require knowledge of the ground truth labels is one of its advantages over other clustering evaluation measures.

Our findings reveal that among the algorithms we examined, spectral clustering had the lowest DBI score of 1.8, indicating the highest clustering performance. Birch clustering and K means both displayed strong performances, with DBI scores of 2.2 and 2.3, respectively. Agglomerative clustering, on the other hand, generated the lowest clustering performance among the investigated algorithms, with a DBI score of 2.6 (Fig. 10).

However, it does have some restrictions. It may be sensitive to factors like the quantity of clusters and the dimensionality of the data, for instance.

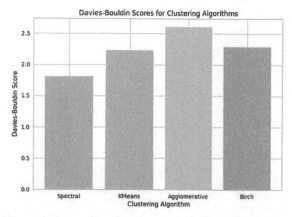

Fig. 10. Shows the comparative analysis of clustering algorithm on the basis of Davies-Bouldin Score.

Overall, the Davies-Bouldin index can be a valuable tool for evaluating the performance of clustering algorithms, but it should be used in conjunction with other metrics and methodologies to have a more thorough knowledge of clustering performance.

5.3 Calinski-Harabasz Score

An additional statistic used to assess the calibre of clustering findings is the Calinski-Harabasz index. A higher score indicates greater clustering ability since it calculates the ratio of between-cluster to within-cluster dispersion. The index is defined as:

$$CH(K) = \frac{\frac{B(K)}{K-1}}{\frac{W(K)}{N-K}} \qquad (2)$$

where K is the number of clusters, B(K) is the between-cluster dispersion, W(K) is the within-cluster dispersion, and N is the total number of data points.

The sum of the squared distances between the centroids of each cluster and the centroid of the entire dataset, multiplied by the quantity of points in each cluster, is the between-cluster dispersion. The total of the squared distances between each data point and its cluster centroid, multiplied by the quantity of points in each cluster, is the within-cluster dispersion.

The best number of clusters is often the one that maximizes the CH score while being sparse. A higher CH score indicates better clustering performance (Fig. 11).

According to our findings, spectral clustering outperformed the other techniques we looked at in terms of clustering performance, having the highest Calinski-Harabasz score 2538. A strong performance was also shown by K means and agglomerative clustering, with Calinski-Harabasz scores of 2080 and 1641, respectively. With a Calinski-Harabasz score of 1573, Birch clustering had a poor clustering performance among the tested techniques.

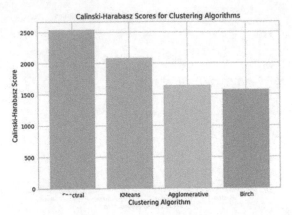

Fig. 11. Shows the comparative analysis of clustering algorithm on the basis of Calinski–Harabasz Score.

6 Conclusion and Future Scope

On the Intel image classification dataset, we tested the effectiveness of the clustering algorithms using three well-known evaluation methods: silhouette score, Calinski-Harabasz index, and Davies-Bouldin index. Overall, the rankings of the clustering algorithms were consistent across the three approaches. With the highest silhouette score, the highest Calinski-Harabasz index, and the lowest Davies-Bouldin index, spectral clustering consistently outperformed the other algorithms in each of the three-evaluation metrics.

In some circumstances, K-means and Agglomerative may also be useful, albeit their performance may be constrained in datasets with overlapping groups. For high-dimensional image data, aggregative clustering might not be the ideal option because it may be less effective at capturing the underlying structure of the data and more susceptible to noise.

Overall, our comparison of these three clustering assessment methodologies indicates that Spectral clustering, followed by K means and Agglomerative clustering, was the most successful algorithm in grouping the Intel picture dataset.

In the future, we plan to solely focus on applying feature extraction techniques to enhance automated scene recognition for environmental monitoring. By exploring various feature extraction methods and their combinations, we aim to optimize the representation of environmental images, leading to improved accuracy and efficiency in scene categorization. Through this focused approach, we intend to contribute to the advancement of automated scene recognition technologies and pave the way for more informed and sustainable environmental monitoring practices.

References

1. Kanungo, T., Mount, D.M., Netanyahu, N.S., Piatko, C., Silverman, R., Wu, A.Y.: The analysis of a simple k-means clustering algorithm. In: Proceedings of the Sixteenth Annual Symposium on Computational Geometry, pp. 100–109, May 2000
2. Deng, D.: DBSCAN clustering algorithm based on density. In: 2020 7th International Forum on Electrical Engineering and Automation (IFEEA), Hefei, China, pp. 949–953 (2020). https://doi.org/10.1109/IFEEA51475.2020.00199
3. Zhang, T., Ramakrishnan, R., Livny, M.: BIRCH: a new data clustering algorithm and its applications. Data Min. Knowl. Disc. **1**, 141–182 (1997)
4. Sasirekha, K., Baby, P.: Agglomerative hierarchical clustering algorithm-a. Int. J. Sci. Res. Publ. **83**(3), 83 (2013)
5. Naeem, A., Rehman, M., Anjum, M., Asif, M.: Development of an efficient hierarchical clustering analysis using an agglomerative clustering algorithm. Curr. Sci. **117**(6), 1045–1053 (2019)
6. Wilkin, G.A., Huang, X.: K-means clustering algorithms: implementation and comparison. In: Second International Multi-Symposiums on Computer and Computational Sciences (IMSCCS 2007), pp. 133–136. IEEE, August 2007
7. Sinaga, K.P., Yang, M.S.: Unsupervised K-means clustering algorithm. IEEE Access **8**, 80716–80727 (2020)
8. Ackermann, M.R., Blomer, J., Kuntze, D., Sohler, C.: Analysis of agglomerative clustering. Algorithmica **69**, 184–215 (2014)
9. Day, W.H., Edelsbrunner, H.: Efficient algorithms for agglomerative hierarchical clustering methods. J. Classif. **1**(1), 7–24 (1984)
10. Zhang, T., Ramakrishnan, R., Livny, M.: BIRCH: an efficient data clustering method for very large databases. ACM SIGMOD Rec. **25**(2), 103–114 (1996)
11. Rahmani, M.K.I., Pal, N., Arora, K.: Clustering of image data using K-means and fuzzy K-means. Int. J. Adv. Comput. Sci. Appl. **5**(7) (2014)
12. Gupta, M.K., Chandra, P.: A comparative study of clustering algorithms. In: 2019 6th International Conference on Computing for Sustainable Global Development (INDIACom), New Delhi, India, pp. 801–805 (2019)
13. Xu, R., WunschII, D.: Survey of clustering algorithms. IEEE Trans. Neural Netw. **16**(3), 645–678 (2005). https://doi.org/10.1109/TNN.2005.845141
14. Murtagh, F.: A survey of recent advances in hierarchical clustering algorithms. Comput. J. **26**(4), 354–359 (1983)
15. Fung, G.: A comprehensive overview of basic clustering algorithms (2001)
16. Ng, A., Jordan, M., Weiss, Y.: On spectral clustering: analysis and an algorithm. Adv. Neural Inf. Process. Syst. **14** (2001)
17. Jia, H., Ding, S., Xu, X., Nie, R.: The latest research progress on spectral clustering. Neural Comput. Appl. **24**, 1477–1486 (2014)
18. Zhang, L., Zhang, L., Du, B.: Deep learning for remote sensing data: a technical tutorial on the state of the art. IEEE Geosci. Remote Sens. Mag. **4**(2), 22–40 (2016)
19. Soudy, M., Afify, Y., Badr, N.: RepConv: a novel architecture for image scene classification on Intel scenes dataset. Int. J. Intel. Comput. Inf. Sci. **22**(2), 63–73 (2022)
20. Wu, Y., Wang, Z., Ripplinger, C.M., Sato, D.: Automated object detection in experimental data using combination of unsupervised and supervised methods. Front. Physiol. **13**, 469 (2022)
21. Zhao, L., Li, S.: Object detection algorithm based on improved YOLOv3. Electronics **9**(3), 537 (2020)
22. Patel, K., Rambach, K., Visentin, T., Rusev, D., Pfeiffer, M., Yang, B.: Deep learning-based object classification on automotive radar spectra. In: 2019 IEEE Radar Conference (RadarConf), pp. 1–6. IEEE, April 2019

Unveiling the Potentials of Deep Learning Techniques for Accurate Alzheimer's Disease Neuro Image Classification

Debahuti Mishra[1($\boxtimes$)], Arundhati Lenka[1], and Sashikala Mishra[2]

[1] Department of Computer Science and Engineering, Siksha 'O' Anusandhan (Deemed to be) University, Bhubaneswar, Odisha, India
{debahutimishra,arundhatilenka}@soa.ac.in
[2] Symbiosis Institute of Technology, Symbiosis International (Deemed University), Pune, Maharashtra, India
sashikala.mishra@sitpune.edu.in

Abstract. This manuscript explores the potential of deep learning strategies, including convolutional neural networks (CNN); recurrent neural network (RNN); long short term memory (LSTM); Bi-directional LTSM (Bi-LSTM), for accurate and automated classification of neuroimages in the diagnosis of Alzheimer's Disease (AD). The study introduces an optimized deep learning model, Bi-LSTM-AJSO, developed using the artificial jellyfish search optimization (AJSO) algorithm. The study conducts a thorough contrast with machine learning and deep learning techniques, along with optimized Bi-LSTM models. The summarized outcomes illustrate the consistent superiority of Bi-LSTM in accuracy, precision, sensitivity, and F1-score. The proposed Bi-LSTM-AJSO stands out among rest of the optimized Bi-LSTM models and surpasses other machine learning and deep learning-based classifiers. Notably, promising convergence and execution time performances are observed for both Bi-LSTM and Bi-LSTM-AJSO. These findings emphasize the effectiveness of Bi-LSTM models and optimized methods in precise AD neuroimage analysis. The research suggests future avenues, including exploring alternate deep learning architectures, and underscores the significance of independently validating datasets for real-world clinical relevance.

Keywords: Alzheimer's Disease Neuro Image Classification · Convolutional Neural Network (CNN) · Recurrent Neural Network (RNN) · Long-short term memory (LSTM) · Bi-directional LTSM (Bi-LSTM) · Artificial Jellyfish Search Optimization (AJSO)

1 Introduction

Alzheimer's disease (AD) is a progressive neurodegenerative disorder causing cognitive decline and memory loss. Accurate early diagnosis is vital for effective management. The disease leads to brain cell loss, especially in memory related regions like the hippocampus. The hippocampus, critical for memory, learning, and navigation, suffers tissue loss as AD advances, severely affecting memory.

S. Tiwari et al. (Eds.): AI4S 2023, CCIS 1907, pp. 74–88, 2023.
https://doi.org/10.1007/978-3-031-47997-7_6

This emphasizes its role in AD pathology and early susceptibility (Fig. 1). Neuroimaging, e.g., magnetic resonance image (MRI), positron emission tomography (PET) etc. reveals AD-related brain changes [1,2]. Yet, manual image interpretation is time-consuming and variable. Machine and deep learning offer a solution by automating neuroimage analysis [3,4]. Deep learning extracts features from raw images, enabling automated analysis, accelerating the process, and ensuring consistency. Learning from extensive data, it improves accuracy, detecting subtle abnormalities overlooked by humans. This boosts neuroimaging diagnosis efficiency and standardization, benefiting clinicians and patients [5,6].

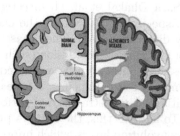

Fig. 1. View of normal brain and AD brain showing massive cell loss in a brain with advanced AD. Tissue loss in the hippocampus is especially severe. The hippocampus is crucial for memory and one of the first areas to be affected by Alzheimer's [7].

The aim of this work is to explore the potential of deep learning strategies for accurate and automated classification of neuroimages to aid in the diagnosis of AD and to provide insights into the neuroanatomical patterns associated with AD using deep learning interpretability techniques. The key contributions of this manuscript are (a) to investigate the use of deep learning techniques such as; convolutional neural network (CNN) [3], recurrent neural network (RNN) [1,2], long-short term memory (LSTM) [8,9], Bi-directional LTSM (Bi-LSTM) [10] for AD neuroimage classifi-cation; (b) to develop a robust and accurate optimized deep learning model based on Bi-LSTM for AD diagnosis using neuroimaging data using a new meta-heuristic artifi-cial jellyfish search optimization (AJSO) algorithm [11,12] coined as Bi-LSTM-AJSO and; (c) to compare the performance of the proposed Bi-LSTM-AJSO model with traditional machine learning and deep learning approaches.

The remainder of this work comprises several sections: a literature survey, methodologies (Sect. 2 and Sect. 3), discussion of the Bi-LSTM-AJSO classification model (Sect. 4), experimentation and result analysis (Sect. 5), and concluding remarks and future directions (Sect. 6).

2   Literature Survey

This discusses the AD detection framework, feature extraction methods, and machine learning algorithms, highlighting the potential of feature extraction.

A different study introduces CNNs for automatic classification of brain scans, specifically focusing on a small region of interest like the hippocampal region. Multiple models are combined for AD diagnosis and classification, and various transfer learning techniques are evaluated, including cross-modal and cross-domain transfer learning.

Arijit De et al. [3] addressed the automated classification of AD and introduced a novel approach to directly classify four classes: AD, normal control, early mild cognitive impairment, and late mild cognitive impairment using 3D diffusion tensor imaging (DTI). DTI provides information about brain anatomy through fractional anisotropy and mean diffusivity values, along with echo planar imaging intensities. Sara Dhakal et al. [4] investigated the application of deep learning techniques, specifically CNNs, for the classification of CT brain images to aid in the early diagnosis of AD. The dataset used in the study consists of three categories of CT images: AD, lesion (such as tumors), and normal aging, totaling 285 images. The authors in [5] evaluated the DL models for AD classification using MR images from Alzheimer's Disease Neuroimage (ADNI) dataset. DenseNet-121 initially demonstrates good performance but suffers from computational slowness. To address this, the authors propose a modified architecture using depth-wise convolution layers, which improves both execution time and classification performance, achieving an average rate of 90.22%. L.V.S.K.B. Kasyap et al. [6] explored advanced deep learning techniques for detecting mental illness. The researchers conducted a comprehensive analysis of papers from authoritative databases, investigating classic machine learning and primary deep learning techniques used for predicting mental health issues.

3   Methodology

This section focuses on implementing deep learning techniques for brain lesion MRI data classification. We explore the effectiveness of CNN [3], RNN [1,2], LSTM [8,9], and Bi-LSTM [10] with AJSO algorithm [11,12]. RNNs are well-suited for analyzing brain MRI sequences as they handle temporal dependencies. They maintain internal memory, capturing long-range dependencies and temporal dynamics. RNNs process sequential MRI images, modelling relationships and extracting features. One popular variant of RNNs is the LSTM network, overcome vanishing gradients by incorporating memory cells with gating mechanisms. This makes LSTMs suitable for brain MRI data with long-term dependencies. Additionally, RNNs can be combined with other neural network architectures, such as CNNs, to leverage their complementary strengths. CNNs are effective in capturing spatial features from individual MRI images, while RNNs excel at capturing temporal dependencies.

The Bi-LSTM network, a specialized variant of LSTM widely used in sequence tasks like brain MRI classification, processes sequences bidirectionally. It considers information from previous and subsequent MRI slices, capturing complex temporal and spatial dependencies. Incorporating both aspects enables it to effectively learn patterns and relationships between slices. Recognizing each slice's unique brain location and variations in time and space enhances overall data understanding. This approach effectively models MRI data dynamics, leading to improved classification. By processing sequential images, the network learns relationships and extracts features, identifying changes, tracking disease progression, and detecting abnormalities. Bi-LSTMs handle variable-length sequences often in brain MRI, adapting dynamically to different protocols or slice counts. This adaptability ensures effective processing of varying sequence lengths. See Fig. 2 for network architecture illustration.

Metaheuristics optimize classifiers by providing effective search strategies, evolving solutions, and considering interactions for complex problems. They improve classifier performance by optimizing hyperparameters, feature selection, and model architecture. This study utilizes the AJSO metaheuristic for optimization. AJSO is inspired by jellyfish behaviour, where they flock to areas with more food. The algorithm follows three rules: (a) Jellyfish movement is influenced by ocean currents or swarm behaviour, controlled by time. (b) Jellyfish actively search for food-rich locations. (c) Jellyfish location depends on food quantity, assessed using an objective function.

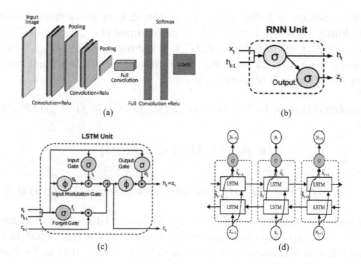

Fig. 2. The architectural representation of CNN, RNN, LSTM and Bi-LSTM networks [1–10].

This study utilizes the AJSO metaheuristic for optimization. AJSO is inspired by jellyfish behaviour, where they flock to areas with more food. The algorithm follows three rules: (a) Jellyfish movement is influenced by ocean currents or swarm behaviour, controlled by time. (b) Jellyfish actively search for food-rich locations. (c) Jellyfish location depends on food quantity, assessed using an objective function. The AJS algorithm incorporates the ocean current, jellyfish swarm, time-controlled mechanism, and boundary conditions. These elements guide the algorithm's behaviour to optimize performance. The jellyfish are guided by the ocean current towards areas with higher food concentrations. The new jellyfish location is determined by Eq. (1) and Eq. (2) [11, 12].

$$Ocean\_Current = JF\_best - \varphi \times rand(0,1) \times MeanLocation\_JF \qquad (1)$$
$$JF\_i(t+1) = JF\_i(t) + rand(0,1) \times Ocean_Current \qquad (2)$$

In the equations, JF_best represents the jellyfish currently at the best location in the swarm or bloom, φ is the distribution coefficient (> 0) related to the direction of the Ocean_Current, JF_i represents the jellyfish with index i, and $MeanLocation\_JF$ represents the new location of each jellyfish. Jellyfish exhibit two types of mobility: passive and active motion. Initially, during the formation of a bloom, jellyfish show passive motion around their own locations. Equation (3) is used to update the location of each jellyfish, where ω is the motion coefficient and ($Upper\_bound - Lower\_bound$) represents the range of movement.

$$JF\_i(t+1) = JF\_i(t) + \omega \times rand(0,1) \times (Upper\_bound - Lower\_bound) \,(3)$$

For active motion, jellyfish either move towards or away from another jellyfish. A randomly chosen jellyfish, JF_j, determines the direction of movement. If the food quantity exceeds that of JF_i, it moves towards JF_j; if it is lower, it moves away. Equation (4) and Eq. (5) update the jellyfish location based on the motion direction and objective function.

$$Motion\ Direction = \{JF\_j(t) - JF\_i(t) if f(JF_i) \geq f(JF\_j)JF\_i(t) - $$

$$JF\_j(t) if f(JF\_i) < f(JF\_j)\} \qquad (4)$$

$$JF\_i(t+1) = JF\_i(t) + rand(0,1) \times (Motion\ Direction) \qquad (5)$$

To control the motions of jellyfish in a bloom, a time-controlled mechanism is used. Equation (6) describes this mechanism using a time control function, $f(T\_C)$, which is determined by a random value between [0,1] and a constant, c. The time control function changes over time based on the iteration number (t) and the maximum number of iterations ($Iterations\_max$).

$$f(T\_C) = |(1 - t/Iterations\_max) \times (2 \times rand(0,1) - 1)| \qquad (6)$$

When the value of $f(T\_C)$ increases and exceeds the constant c, it indicates that the jellyfish follow the ocean current (Ocean_Current). On the other hand, when $f(T\_C)$ is less than c, the jellyfish move within the bloom. The specific value of c when $f(T\_C)$ equals c is not known as it changes over time. The boundary conditions ensure that jellyfish stay within the defined search space. Equation (7) describes the behavior of a jellyfish when it reaches the boundaries of the search area. If the location of the ith jellyfish in the dth dimension, represented as $JF(i,d)$, exceeds the upper bound $Upper(bound, d)$, the jellyfish's new location with respect to Eq. (7) and on the other hand, if $JF(i, d)$ falls below the lower bound $Lower(bound, d)$, the new location is obtained using Eq. (8).

$$JF(i, d)' \text{ is calculated as } (JF(i, d) - Upper\_bound, d) + Lower\_bound(d) \quad (7)$$

$$JF(i, d)' \text{ is calculated as } (JF(i, d) - Lower\_bound, d) + Upper\_bound(d) \quad (8)$$

This ensures that the jellyfish remains within the search space by wrapping around to the opposite bound when it reaches a boundary. By combining the AJSO algorithm's population-based search strategy with the Bi-LSTM network, this ap-proach aims to discover optimal or near-optimal hyperparameter configurations that maximize the network's performance in brain MRI data classification. The iterative nature of AJSO allows for an efficient exploration of the hyperparameter space, lead-ing to improved classification accuracy and better understanding of neurological conditions through accurate brain abnormality detection.

4 The Proposed Bi-LSTM-AJSO Model Development

In this section, we employ deep learning techniques for brain lesion MRI classification using Bi-LSTM networks with the AJSO algorithm inspired by jellyfish swarms. AJSO optimizes Bi-LSTM network parameters by iteratively exploring the search space, refining hyperparameters such as LSTM layers, hidden units, learning rate, and dropout rate. AJSO leverages swarm intelligence for efficient near-optimal configuration discovery. Combining Bi-LSTM's capacity for capturing sequential and spatial dependencies with AJSO's parameter tuning enhances MRI classification accuracy and robustness. The model's layout, experimentation stages, and comparison steps are illustrated in Fig. 3.

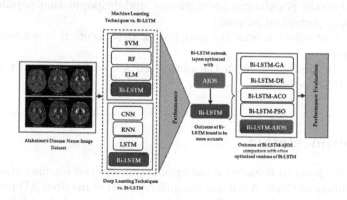

Fig. 3. Schematic outline of the proposed Bi-LSTM-AJSO brain lesion MRI classification of AD.

By iteratively applying the steps outlined in Algorithm 1, the AJSO aims to optimize the hyperparameters of the Bi-LSTM network for brain MRI data classification, enhancing the network's performance in accurately identifying brain abnormalities and improving the understanding of neurological conditions.

Algorithm 1. LSTM-AJSO Classification

Inputs: population_size (integer): The size of the population max_iterations (integer): The maxi-mum number of iterations termination_threshold (float): The threshold for termination

Output: best_solution ( ): The best set of hyperparameters for the Bi-LSTM network

---

1. Initialize population: population = initialize_population(population_size)
2. Iterate for each iteration in the range from 1 to max_iterations: for iteration = 1 to max_iterations:
   2.1 Evaluate fitness for each jellyfish in the population: for each jellyfish in population: fitness = evaluate_fitness(jellyfish) jellyfish.fitness = fitness
   2.2 Get the best jellyfish in the population: best_jellyfish = get_best_jellyfish (population)
   2.3 Check if the fitness of the best jellyfish meets the termination threshold: if best_jellyfish.fitness = termination_threshold: break
   2.4 Perform selection to choose individuals for reproduction: selected_jellyfish = selection(population)
   2.5 Perform crossover to create offspring: offspring = crossover(selected_jelly fish)
   2.6 Perform local search to perturb a subset of jellyfish individuals: per-turbed_jellyfish = local_search(selected_jellyfish)
   2.7 Update the population: population = update_population(population, off-spring, perturbed_jellyfish)
3. Get the best solution from the population: best_solution = get_best_jellyfish (population)
4. Return the best_solution as the optimized set of hyperparameters for the Bi-LSTM network.
5. End

---

5 Experimentation

This section aimed to investigate the application of deep learning strategies for AD neuroimage analysis. A dataset comprising MRI scans from AD patients and healthy controls was collected. Deep learning models, including CNNs, RNNs, LSTM and Bi-LSTM, were trained to extract relevant features from the neuroimages and learn spatial patterns. The experimental evaluation involved AD classification, comparing the performance of our optimized deep learning approach with traditional machine learning and deep learning methods. The experimentation has been performed using Intel(R) Core(TM) i5-7200U CPU @ 2.50

GHz with 2.71 GHz processor, 4.00 GB (3.88 GB usable) RAM, 64-bit operating system, x64-based processor operating sys-tem and executed on the platform Google Colab using python libraries namely Keras, Numpy, Panda and Sklearn.

5.1   Datasets Used and Model Training and Testing

In this research work, the ADNI [13] dataset has been used for experimentation which consists of 360 samples from individuals with normal cognitive function and 290 samples from individuals diagnosed with AD. In total, there are 455 samples available in the dataset. Out of these, 195 samples are allocated for training purposes, while the remaining 260 samples are designated for testing. This dataset contains a diverse range of brain images, allowing for the development and evaluation of classification models for different brain conditions. The sample of scan slices of the ADNI dataset is given in Fig. 4.

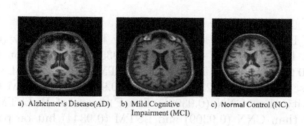

a) Alzheimer's Disease(AD)   b) Mild Cognitive   c) Normal Control (NC)
                             Impairment (MCI)

Fig. 4. Sample scan slices from the ADNI dataset [13]

5.2   Comparison with Other Machine and Deep Learning Approaches

This empirical study presents a comprehensive comparison of Bi-LSTM with traditional machine learning methods, including support vector machines (SVM), random forest (RF), and extreme learning machines (ELM), as well as with other deep learning-based approaches [3–10]. Furthermore, the study compares different variations of Bi-LSTM, with genetic algorithm (GA), differential evolution (DE), particle swarm optimization (PSO) and ant colony optimization (ACO) [14] such as Bi-LSTM-AJOS, Bi-LSTM-GA, Bi-LSTM-DE, Bi-LSTM-PSO, and Bi-LSTM-ACO. The evaluation of these models is based on metrics such as accuracy, precision, sensitivity, and F1-score. Additionally, the study discusses the overall improvements achieved by Bi-LSTM across all three categories of classifiers.

Bi-LSTM demonstrates superior performance compared to other classifiers in Table 1, exhibiting higher accuracy, precision, sensitivity, and F1-score. It achieves an accuracy of 0.9357, surpassing SVM (0.9125), RF (0.9110), and ELM (0.9102). Bi-LSTM also achieves higher precision (0.9358) compared to SVM (0.9124), RF (0.9102), and ELM (0.9100). Its sensitivity (0.9322) outperforms SVM (0.9024), RF (0.9108), and ELM (0.9008), indicating superior identification of positive instances. Furthermore, Bi-LSTM's F1-score (0.9324) is higher than SVM (0.9155), RF (0.9058), and ELM (0.9024), demonstrating a balanced metric that accounts for precision and sensitivity. These results high-light Bi-LSTM's superior performance compared to tra-ditional machine learning classifiers, showcasing its ability to capture long-term dependencies.

Table 1. Performance comparison of Bi-LSTM with traditional machine learning based-classifiers.

Classifiers	Accuracy	Precision	Sensitivity	F1-Score
SVM	0.9125	0.9124	0.9024	0.9155
RF	0.9110	0.9102	0.9108	0.9058
ELM	0.9102	0.9100	0.9008	0.9024
Bi-LSTM	**0.9357**	**0.9358**	**0.9322**	**0.9324**

In Table 2, Bi-LSTM also achieves slightly higher accuracy (0.9357) than CNN (0.9322), RNN (0.9356), and LSTM (0.9355). Its precision (0.9358) sur-passes CNN (0.9321), RNN (0.9355), and LSTM (0.9354). Bi-LSTM exhibits slightly lower sensitivity (0.9322) than RNN (0.9354) and LSTM (0.9348), but higher sensitivity than CNN (0.9302). The F1-score of Bi-LSTM (0.9324) is slightly higher than CNN (0.9309) and LSTM (0.9344) but on par with RNN (0.9344). Overall, Bi-LSTM con-sistently outperforms the deep learning classi-fiers used for comparison.

Table 2. Performance comparison of Bi-LSTM with traditional deep learning-based classifiers

Classifiers	Accuracy	Precision	Sensitivity	F1-Score
CNN	0.9322	0.9321	0.9302	0.9309
RNN	0.9356	0.9355	0.9354	0.9344
LSTM	0.9355	0.9354	0.9348	0.9344
Bi-LSTM	0.9357	0.9358	0.9322	0.9324

In Table 2, Bi-LSTM also achieves slightly higher accuracy (0.9357) than CNN (0.9322), RNN (0.9356), and LSTM (0.9355). Its precision (0.9358) sur-passes CNN (0.9321), RNN (0.9355), and LSTM (0.9354). Bi-LSTM exhibits slightly lower sensitivity (0.9322) than RNN (0.9354) and LSTM (0.9348), but higher sensitivity than CNN (0.9302). The F1-score of Bi-LSTM (0.9324) is slightly higher than CNN (0.9309) and LSTM (0.9344) but on par with RNN (0.9344). Overall, Bi-LSTM con-sistently outperforms the deep learning classi-fiers used for comparison.

Table 3. Performance comparison of Bi-LSTM-AJOS with other optimized versions of Bi-LSTM classi-fiers

Classifiers	Accuracy	Precision	Sensitivity	F1-Score
Bi-LSTM-GA	0.9325	0.9311	0.9301	0.9289
Bi-LSTM-DE	0.9511	0.9502	0.9601	0.9588
Bi-LSTM-PSO	0.9655	0.9614	0.9598	0.9603
Bi-LSTM-ACO	0.9754	0.9722	0.9735	0.9768
Bi-LSTM-AJOS	0.9876	0.9855	**0.9889**	**0.9899**

Table 3 compares the performance of Bi-LSTM-AJOS with other optimized versions of Bi-LSTM classifiers: Bi-LSTM-GA, Bi-LSTM-DE, Bi-LSTM-PSO, and Bi-LSTM-ACO. Bi-LSTM-GA achieves an accuracy of 0.9325, while Bi-LSTM-DE demonstrates significant improvement with an accuracy of 0.9511. Bi-LSTM-PSO further improves performance, achieving an accuracy of 0.9655. Notably, Bi-LSTM-ACO achieves high accuracy (0.9754). However, Bi-LSTM-AJOS surpasses all other classifiers, achieving an exceptional accuracy of 0.9876. These results highlight Bi-LSTM-AJOS' effectiveness in accurate classification, outperforming the other opti-mized versions across all evaluation metrics.

Table 4 compares the performance of Bi-LSTM and Bi-LSTM-AJOS with other machine learning-based classifiers: Bi-LSTM-GA, Bi-LSTM-DE, Bi-LSTM-PSO, and Bi-LSTM-ACO. The evaluated metrics are accuracy, precision, sensitivity, and F1-score. Bi-LSTM-GA achieves an accuracy of 0.9325, while Bi-LSTM-DE shows significant improvements across all metrics, achieving an accuracy of 0.9511. Further improvements are observed in Bi-LSTM-PSO, with an accuracy of 0.9655, and Bi-LSTM-ACO, with an accuracy of 0.9754. The optimized version, Bi-LSTM-AJOS, surpasses all classifiers with exceptional performance, achieving an accuracy of 0.9876. These results demonstrate the effectiveness of the optimization techniques implemented in the respective models, leading to superior classification performance.

Table 4. Accuracy improvement of Bi-LSTM-AJSO over machine learning based-classifiers

Bi-LSTM vs. SVM	Bi-LSTM vs. RF	Bi-LSTM vs. ELM	Bi-LSTM-AJOS vs. SVM	Bi-LSTM-AJOS vs. RF	Bi-LSTM-AJOS vs. ELM
Accuracy					
2.4794271	2.6397349	2.7252324	7.6042932	7.7561765	7.8371810
Precision					
2.500534	2.7356272	2.7569993	7.4175545	7.6407914	7.6610857
Sensitivity					
3.196738	2.2956447	3.3683758	8.7470927	7.8976640	8.9088886
F1-Score					
1.8125268	2.8528528	3.2175032	7.515910	8.4958076	8.8392766

Table 5 provides a comparison of the performance of Bi-LSTM-AJOS with various deep learning-based classifiers and optimized versions of Bi-LSTM: CNN, RNN, LSTM, Bi-LSTM, Bi-LSTM-GA, Bi-LSTM-PSO, Bi-LSTM-DE, and Bi-LSTM-ACO. The metrics evaluated are accuracy, precision, sensitivity, and F1-score. Comparing Bi-LSTM-AJOS with CNN, RNN, and LSTM, it achieves higher accuracy, precision, sensitivity, and F1-score values. This indicates that Bi-LSTM-AJOS outper-forms these deep learning classifiers in accurately classifying instances. In comparison with the optimized versions of Bi-LSTM, Bi-LSTM-AJOS demonstrates superior performance in terms of accuracy, precision, sensitivity, and F1-score. It surpasses Bi-LSTM-GA, Bi-LSTM-PSO, Bi-LSTM-DE, and Bi-LSTM-ACO, highlighting the effectiveness of the artificial immune optimization technique implemented in Bi-LSTM-AJOS.

Table 5. Accuracy improvement of Bi-LSTM-AJSO over other optimized versions of Bi-LSTM classifiers

Bi-LSTM AJOS vs. CN	Bi-LSTM-AJOS vs. RNN	Bi-LSTM-AJOS vs. LSTM	Bi-LSTM-AJOS vs. Bi-LSTM	Bi-LSTM-AJOS vs. Bi-LSTM-GA	Bi-LSTM-AJOS vs. Bi-LSTM-PSO	Bi-LSTM-AJOS vs. Bi-LSTM-DE	Bi-LSTM-AJOS vs. Bi-LSTM-ACO
Accuracy							
5.6095585	5.2652895	5.2754151	5.2551640	5.5791818	3.6958282	2.2377480	1.2353179
Precision							
5.4185692	5.0735667	5.0837138	5.0431253	5.5200405	3.5819381	2.4454591	1.3495687
Sensitivity							
5.9358883	5.4100515	5.4707250	5.7336434	5.9460006	2.9123268	2.9426635	1.5572858
F1-Score							
5.960198	5.6066269	5.6066269	5.8086675	6.1622386	3.1417314	2.9902010	1.323366

The results presented in Table 5 indicate that Bi-LSTM-AJOS performs exceptionally well compared to both deep learning-based classifiers and optimized versions of Bi-LSTM. It achieves higher accuracy and exhibits improvements in precision, sensitivity, and F1-score. These findings suggest that Bi-LSTM-AJOS is highly effective in accurately classifying instances and outperforms other models in the evaluation metrics.

5.3   Execution Time Comparisons

The convergence curve represents the behavior of the mean squared error (MSE) values as the iterations progress during the training process. A steeper decrease in the MSE indicates faster convergence and better performance of the model. In the analysis of convergence curves measured based on the MSE of various optimized models of Bi-LSTM, including Bi-LSTM-GA, Bi-LSTM-DE, Bi-LSTM-PSO, Bi-LSTM-ACO, and the proposed Bi-LSTM-AJOS, it was observed that the Bi-LSTM-AJOS converged quickly around the 45th iteration and outperformed the rest of the compared optimized models as shown in Fig. 5.

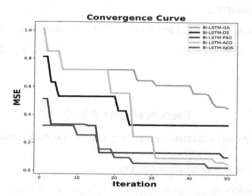

Fig. 5. Schematic outline of the proposed Bi-LSTM-AJSO brain lesion MRI classification of AD.

In this case, the Bi-LSTM-AJOS exhibited a significant improvement in convergence speed compared to the other models. By reaching convergence quickly, the Bi-LSTM-AJOS demonstrated its ability to effectively optimize the weights and biases of the Bi-LSTM network. Furthermore, the Bi-LSTM-AJOS not only converged faster but also exhibited better performance in terms of MSE. Lower MSE values indicate better accuracy and predictive capability of the model. The improved convergence speed of Bi-LSTM-AJOS potentially allowed it to find more optimal solutions, resulting in lower MSE values compared to the other models. Similarly, the execution time (measured in minutes) performance of the Bi-LSTM model with respect to traditional machine learning, deep learning based and optimized Bi-LSTM are compared as shown in Fig. 6(a), Fig. 6(b), and Fig. 6(c) respectively.

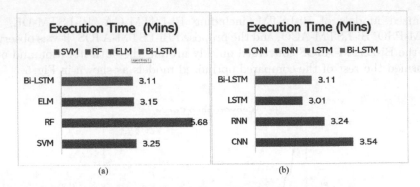

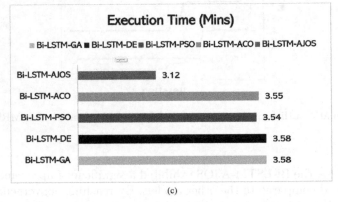

Fig. 6. Execution time performance evaluation of (a) Bi-LSTM vs. machine learning based classifiers; (b) Bi-LSTM vs. deep learning-based classifiers and (c) Bi-LSTM-AJOS vs. Bi-LSTM-GA, Bi-LSTM-DE, Bi-LSTM-PSO, Bi-LSTM-ACO.

5.4   Interpretability Analysis and Significance

The summarized results provide an overview of the performance and effectiveness of Bi-LSTM and Bi-LSTM-AJOS in various comparisons and evaluations. Here are the key insights from the interpretation and analysis:

(a) Bi-LSTM consistently outperforms other classifiers in accuracy, precision, sensitivity, and F1-score, showcasing its ability to capture long-term dependencies and establish itself as a highly capable model (Table 1 and Table 2).

(b) Bi-LSTM-AJOS achieves exceptional performance among optimized Bi-LSTM classifiers, exhibiting high accuracy, precision, sensitivity, and F1-score (Table 3). Incorporating optimization techniques enhances the effectiveness of Bi-LSTM classifiers.

(c) Both Bi-LSTM and Bi-LSTM-AJOS outperform other machine learning-based classifiers in accuracy, precision, sensitivity, and F1-score (Table 4). Optimization techniques further enhance the performance of Bi-LSTM models, making them highly effective for classification tasks.

(d) Bi-LSTM-AJOS surpasses deep learning-based classifiers and optimized versions of Bi-LSTM in accuracy, precision, sensitivity, and F1-score (Table 5). Its exceptional performance highlights its effectiveness in accurately classifying instances.

(e) Bi-LSTM's and Bi-LSTM-AJOS's convergence performance and execution time performance are also showing promising result (Fig. 5 and Fig. 6).

This study's findings pave the way for further research in AD prognosis, disease progression monitoring, and treatment response prediction. Exploring other deep learning architectures like transformers or graph neural networks could yield additional insights, potentially enhancing classification performance and deepening AD understanding. Combining multiple neuroimaging modalities, including structural MRI, functional MRI, and PET scans, can boost classification accuracy and enrich AD comprehension. Integrating these modalities allows a comprehensive disease evaluation, shedding light on underlying pathology. Independent dataset validation is crucial to ensure model generalizability in real-world clinical settings, fortifying findings' robustness and practical usability. In conclusion, utilizing proposed deep learning models, like Bi-LSTM-AJSO, shows promise for real-world AD applications. This research not only improves neuroimage classification but also sets the stage for future AD research and clinical advancements.

6   Conclusion and Future Directions

This proposal investigates deep learning techniques (LSTM, RNN, CNN, and Bi-LSTM) for precise automated AD neuroimage classification. Contributions encompass exploring these techniques, creating an optimized Bi-LSTM model using AJSO (Bi-LSTM-AJSO), and comparing its performance against traditional methods. Bi-LSTM-AJSO excels, surpassing other models. These findings underscore Bi-LSTM's value for accurate AD neuroimage analysis. The study encourages future research in AD prognosis, progression monitoring, and treatment prediction using deep learning models. Examining transformers or graph neural networks may enhance insights and performance. Integrating diverse neuroimaging modalities (structural MRI, functional MRI, PET scans) could improve accuracy and understanding. Independent dataset validation ensures model applicability in real clinical settings.

References

1. Niyas, K.P.M., Thiyagarajan, P.: A systematic review on early prediction of Mild cognitive impairment to Alzheimers using machine learning algorithms. Int. J. Intell. Netw. **4**, 74–88 (2023)
2. Garg, N., Choudhry, M.S., Bodade, R.M.: A review on Alzheimer's disease classification from normal controls and mild cognitive impairment using structural MR images. J. Neurosci. Methods **384**, 109745 (2023)

3. De, A., Chowdhury, A.S.: DTI based Alzheimer's disease classification with rank modulated fusion of CNNs and random forest. Expert Syst. Appl. **169**, 114338 (2021)
4. Dhakal, S., Azam, S., Hasib, K.M., Karim, A., Jonkman, M., Al Haque, A.S.M.F.: Dementia prediction using machine learning. Procedia Comput. Sci. **219**, 1297–1308 (2023)
5. Hazarika, R.A., Kandar, D., Maji, A.K.: An experimental analysis of different deep learning based models for Alzheimer's disease classification using brain magnetic resonance images. J. King Saud Univ. Comput. Inf. Sci. Part A **34**(10), 8576–8598 (2022)
6. Varanasi, L.K., Dasari, C.M.: Deep learning based techniques for Neuro-degenerative disorders detection. Eng. Appl. Artif. Intell. **122**, 106103 (2023)
7. https://www.dovemed.com/health-topics/focused-health-topics/what-alzheimers-disease/ . Accessed 06 June 2023
8. Gill, H.S., Khehra, B.S.: An integrated approach using CNN-RNN-LSTM for classification of fruit images. Mater. Today Proc. Part 1 **51**, 591–595 (2022)
9. Chowdhury, S.R., Khare, Y., Mazumdar, S.: Classification of diseases from CT images using LSTM-based CNN. In: Polat, K., Öztürk, S. (eds.) In Intelligent Data-Centric Systems, Diagnostic Biomedical Signal and Image Processing Applications with Deep Learning Methods, pp. 235–249. Academic Press (2023)
10. Karnam, N.K., Dubey, S.R., Turlapaty, A.C., Gokaraju, B.: EMGHandNet: a hybrid CNN and Bi-LSTM architecture for hand activity classification using surface EMG signals. Biocybern. Biomed. Eng. **42**(1), 325–340 (2022)
11. Chou, J.-S., Truong, D.-N.: A novel metaheuristic optimizer inspired by behavior of jellyfish in ocean. Appl. Math. Comput. **389**, 125535 (2021)
12. Khare, A., Kakandikar, G.M., Kulkarni, O.K.: An insight review on jellyfish optimization algorithm and its application in engineering. Rev. Comput. Eng. Stud. **9**(1), 31–40 (2021)
13. https://adni.loni.usc.edu/ . Accessed 06 Mar 2023
14. Fister, I., Yang, X.-S., Fister, I., Brest, J., Fister, D.: A Brief Review of Nature-Inspired Algorithms for Optimization. arXiv abs/1307.4186 (2013)

Food Composition Knowledge Extraction from Scientific Literature

Azanzi Jiomekong[1](✉), Martins Folefac[2], and Hippolyte Tapamo[1]

[1] Department of Computer Science, University of Yaounde 1, Yaounde, Cameroon
fidel.jiomekong@facsciences-uy1.cm
[2] neuralearn.ai, Yaounde, Cameroon

Abstract. Adequate nutrition is an essential catalyst for economic and human development as well as for achieving Sustainable Development Goals (Goal 2 and Goal 3). Thus, understanding Food Composition (stored in Food Composition Tables) can allow people to have a healthy diet and avoid overnutrition and undernutrition which are cause of lots of health problems. Food Composition Tables (FCT) or Food Composition Databases (FCD) contains the food we eat and what it contains. It is built by using chemical analysis to determine the different composition and structure of foods. However, the chemical analysis of food requires significant financial resources and skilled laboratory investigators. These resources are not always available. Given that many FCT are stored in scientific papers related to food, nutrition, food chemistry, etc. in the form of tables, we propose in this paper to extract these knowledge for the purpose of building Food Composition Tables. The latter can therefore be used for food recommendation, ingredient substitution, etc. To demonstrate the relevance of the knowledge extracted, we invited one domain expert for validation. On the other hand, we compared an excerpt of the knowledge extracted to several biomedical ontologies, food ontologies and we matched some elements extracted to FoodOn ontology and Wikidata Knowledge Graph.

Keywords: Scientific literature · Food and food components · Food Composition Tables · Knowledge Extraction · Ontology learning · Learning Food Composition Knowledge

1 Introduction

Adequate nutrition is an essential catalyst for economic and human development as well as for achieving Sustainable Development Goals [28] - Goal 2: Zero Hunger and Goal 3: Ensure healthy lives and promote well-being for all at all ages. Thus, food information engineering aims to gather, process and furnish food information to stockholders for decision making and action [14]. To this end, AI technologies are used for many purposes such as food recognition and nutrition assessment [22,29], measuring the environmental impact of food production [33], etc. This paper is about the extraction of Food Composition Tables from scientific papers.

© The Author(s), under exclusive license to Springer Nature Switzerland AG 2023
S. Tiwari et al. (Eds.): AI4S 2023, CCIS 1907, pp. 89–103, 2023.
https://doi.org/10.1007/978-3-031-47997-7_7

Food Composition Tables (FCT) or Food Composition Databases (FCD) [5] are used for compiling data about the food we eat and what it contains. Thereafter, these databases are used for a variety of purposes such as clinical practices, public health/education and nutrition monitoring, food industry, food regulation, research, etc. Understanding Food Composition stored in Food Composition Tables can allow people to have a healthy diet and avoid a lot of health problems such as diabetes, anemia, stroke, etc. In the rest of this paper, we'll be using FCT to designate both FCT and FCD.

Food Composition Tables are constructed using chemical analysis of foods [5]. The latter studies the different composition and structure of foods. The qualitative analysis of a food can be used to determine if a substance is present in this food and the quantitative analysis is used to find out the quantity (in numbers with unit) of a substance in a food. However, the chemical analytical method requires significant financial resources and skilled laboratory investigators. This makes the construction of FCT costly in time and resources [5,19]. It should be noted that in many cases, people who are in need of FCT don't have these resources. Thus, a solution consists of relying on existing resources such as scientific literature.

Acquiring food knowledge from scientific papers manually is costly in time and resources, and not scalable. In effect, scientific papers are scattered on the Internet, and one has to search them. Once found, Food Composition Knowledge are manually extracted from them by copying the elements one by one to build Food Composition Tables [19,20,30]. Manual information extraction from a large number of data sources is a cumbersome task and time consuming [15]. In the context of extracting food composition knowledge from scientific papers, this consists of reading the paper one by one, copying the tables from them to build the new system [19]. In the case when we have thousands of papers, this task will need a lot of resources and be time consuming. In this paper, we propose the automatic extraction of Food Composition knowledge from tables stored in scientific papers. This work is part of the TSOTSA project[1] which aims to provide resources such as methodologies, methods, tools, datasets for food information engineering. The direct application of this work is the building of a large scale Food Composition Table dataset[2].

In the rest of this paper, we present firstly Food Composition Knowledge in Sect. 2. Thereafter, we present Food Composition knowledge extraction in Sect. 3, knowledge validation in Sect. 4 and we conclude in Sect. 5.

2   Food Composition Knowledge

Food Composition Knowledge involves all the food entities, properties and relations that can be found in Food Composition Tables. In a recent work [13,16,17], we extracted and compared several key-insights that are used to describe food

---

[1] https://github.com/jiofidelus/tsotsa.
[2] https://github.com/jiofidelus/tsotsa/tree/main/TSOTSATable_dataset.

and their composition. These works showed that Food Composition Knowledge can be organized as follow:

- **Entities:** these are all the food elements that exist. In this work, we found the following entities that can be identified as Food Composition knowledge:
  - **Food:** These are all the substances (from natural or industrial origin) that are provided for human consumption. Examples involve potatoes, rice, banana, orange, apple, yogurt, donut, red meat, seafood, poultry, pizza, Eru, etc.
  - **Food component:** the food we eat contains a certain number of food components such as water, carbohydrates, proteins, fats, vitamins, minerals, dietary fiber, etc.
  - **Food group:** given that many foods share similar properties such as food components, they can be organized into food groups. For instance, given that fruit composition is based on a dry or a fresh weight basis, we can consider fruit as a food group.
  - **Food ingredient:** These are foods that are composed to obtain new foods. It should be noted that a food can also be a food component.
- **Properties:** these are used to describe the characteristics of the food entities presented above. They are composed of DataProperties and ObjectProperties.
  - DataProperties are properties whose values are data types. For instance, *"quantity of Carbohydrates"* of type *"Number"* can be a property used to describe a food component; the food color or smelling of type *"String"* can be used to describe a food of type.
  - ObjectProperties are special attributes whose values are individuals of entities. In effect, the food entities presented above are related to each other. For instance, there is a relation between food and food group (a food belongs to a food group); food and food component (a food contains one or many food components); etc. Thus, *"contain"* can be used to define a relationship between the entity *"Food"* and the entity **"Food component"**. These relations can allow us to obtain food thesaurus.
- **Taxonomy:** this is one of the most important relations used to organize food knowledge. It is used to organize entities and properties through which inheritance mechanisms can be applied. For instance, if we consider the example of fruit presented above, we can consider that there are two fruit sub-groups: fleshy fruits and dry fruits. Thus, by determining the difference between fruits, we can build a taxonomy of fruits.

3 Food Composition Knowledge Extraction from Scientific Papers

This section presents how knowledge sources were identified (Sect. 3.1) and machine learning models used for extracting knowledge from these knowledge sources (Sect. 3.2).

3.1   Knowledge Sources

In recent works, Food Composition Knowledge were extracted from scientific papers to build a large-scale corpus of Food Composition Tables [16,17]. In this section, we present how we used to identify and download the scientific papers used.

We started searching papers in July 2022 and around 5000 papers were downloaded, some papers were found using Google Search Engine and other ones were directly downloaded from INFOOD [4], Langual [11], Semantic scholar[3], Scopus, Journal of Food Composition and Analysis and Food Chemistry Journal.

At the beginning of this work, we were considering only the scientific supports found on the INFOOD[4] and LanguaL[5] repositories because these platforms are used to compile and organize resources describing foods eating in the world. The study of 28 PDF documents downloaded showed that they were outdated. This constraint forced us to consider using Google Search Engine to find more up-to-date papers. Google Search was firstly used to get papers/documents that are related to our topic. Given the large number of results furnished by Google (which was not easy to explore) and the small number of relevant documents (around 30% found after browsing 10 first pages), we decided to use more specialized repositories. Thus, after we downloaded relevant papers from the first 10 pages, we decided to consider scientific papers repositories such as Semantic Scholar.

We used the keywords "Food Composition", "Food Composition Table", "Food Composition Database" to search for papers and we found 4,590,000 - 170,000 - 154,000 papers respectively. The analysis of the first results pages using the papers metadata allowed us to remark that "Food Composition Table" and "Food Composition Databases" were giving more reliable results than "Food Composition". We decided to consider papers furnished with the keyword "Food Composition Tables" and to subscribe to the Semantic Scholar feed in order to be alert to all the papers related to the three keywords. Exploiting the papers title, abstract and keywords we determined the ones that are related to FCT and 52 papers were selected. However, at the date we were searching papers from Semantic Scholar, papers like Food Composition Morocco [19] found using Google Search were not indexed. Thus, we decided to consider the Scopus repository.

Searching papers on the Scopus scientific repository using the "Food Composition", "Food Composition Table", and "Food Composition Database" keywords provided 146,125 - 2,221 - 3,241 papers respectively. Scopus provides us with an interface summarizing the results according to many criteria such as year of publication, subject areas, document types, source title, countries, source types, language of publication, etc. On the other hand, given, that these criteria can allow us to select relevant papers to our subject, we decided to use the keyword "Food Composition" and from "subject areas", "source title", "source

---

[3] https://www.semanticscholar.org/.

[4] https://www.fao.org/infoods/infoods.

[5] https://www.langual.org.

types", select only documents related to "Food", "Food Composition", "Nutrient". From the papers retrieved, we removed all documents of the following types: "Notes", "Editorial", "Letter", "Short Survey", "Erratum", "Conference Review", "Retracted", "Abstract Report", "Undefined". We selected the following documents: "Article", "Review", "Conference Paper", "Book Chapter", "Book", "Data Paper", "Report". We remarked during the use of Scopus download manager that for many editors the papers should be downloaded from the journal wizard. That's why we consider searching for additional papers in two specific journals of the domain: Journal of Food Composition and Analysis and Journal of Food Chemistry.

Using the science direct[6] search engine, the searches of papers using the keywords "Food Composition Tables" in Journal of Food Chemistry gave 3,014 papers. After browsing 500 first papers we remarked that many were not containing tables related to Food and Food Composition. From the Journal of Food Composition and Analysis, we found 3,533 using the keywords "Food Composition Table". From the two journals, we selected only papers of type "Review articles", "Research articles", and removed papers of type "Conference abstracts", "Book review", "Correspondence", "Errata", "Short communications" and "Other". In total, 220 papers were downloaded from Journal of Food Chemistry and 3152 papers from Journal of Food Composition and analysis.

3.2   Knowledge Extraction

Learning food knowledge from tables stored in scientific papers relied on the process presented by the Fig. 1. It consists of table detection, text detection, text recognition, table reconstruction. Concerning the implementations, we relied on PaddleOCR[7] [21] implementations of the different layout detection, text detection and text recognition algorithms. In effect, PaddleOCR allows the detection and recognition of text, it supports more than 80 kinds of multi-language recognition models, semi-automatic data annotation tool (PPOCRLabel) annotating rectangular boxes, irregular texts, table and key information annotation modes, etc.

PaddleOCR was trained with the Paddle framework in the Python programming language. Here the detection and recognition algorithms are neural networks which are trained on a large corpus of data. The source code is freely available[8] with a quickly online demo on Google Colaboratory[9] and a video that can be used for table extraction from scientific papers[10].

To train the table detection, text detection and text recognition algorithms, the following datasets were used:

---

[6] https://www.sciencedirect.com/search.
[7] https://pypi.org/project/paddleocr/.
[8] https://github.com/Neuralearn/pdf-to-excel.
[9] https://colab.research.google.com/drive/1gOPBCVO9VtKcoIewXyr_6nNoxo1Bk
    qbz.
[10] https://youtu.be/HZh31OGiQRQ.

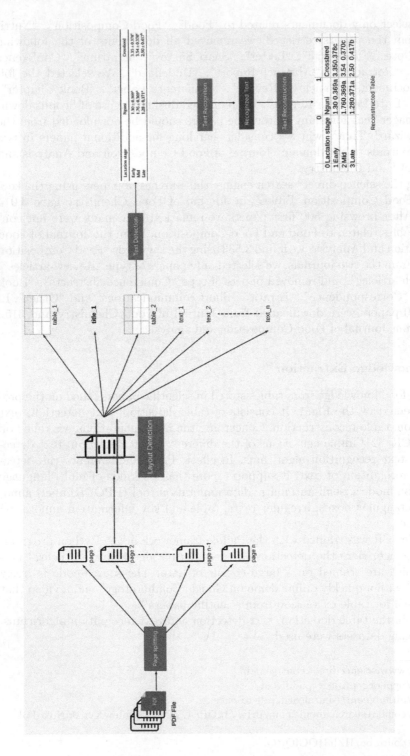

Fig. 1. Pipeline presenting how Food Composition Knowledge are learned from scientific papers

- PubLayNet[11] [35]: this is a large dataset developed to support the development and evaluation of models for document layout analysis.
- Large-scale Street View Text (LSVT) [27]: this dataset contains images captured from streets, which consist of a large variety of complicated real-world scenarios.
- Reading Chinese Text in the Wild - 17 (RCTW-17) [26]: this is a large-scale natural images collected by cameras or mobile phones and in which text instances are annotated with labels, fonts, languages, etc.
- MTWI 2018 [7]: this dataset is a large-scale web images based dataset covering diverse types of web text, including multi-oriented text, tightly-stacked text, and complex-shaped text.
- CASIA-10K [8]: CASIA-10K is a Chinese scene text dataset containing images of various scenarios.
- Scanned receipts OCR and information extraction (SROIE) [10]: this is a large-scale and well-annotated invoice dataset.
- Multi-Lingual scene Text-2019 (MLT-2019) [23]: this is a Multi-Lingual scene Text detection and recognition dataset.
- Born Digital Image (BDI) [18]: BDI dataset is composed of images from five standard word image datasets using semi-automated segmentation.
- MSRA Text Detection 500 (MSRA-TD500) [34]: this dataset contains fully annotated natural images, which are taken from indoor (office and mall) and outdoor (street) scenes using a pocket camera.
- Chinese City Parking Dataset (CCPD) [32]: this is a large-scale dataset for license plate detection and recognition.

To detect the tables from the PDF documents, the first step consists of splitting the document into different pages according to the number of pages of the document. Thereafter, for each page, the document layout analysis [24] is performed to detect tables. The document layout analysis consists of separating a given page into several parts corresponding to the elements (tables, cells, text, pictures, list, etc.) in this page.

The training algorithm used in this work is based on Neural Networks, particularly Deep Convolutional Neural Network. We used PaddlePaddle You Only Look Once (PP-YOLOv2) [9] with a ResNet (Deep Residual Learning for Image Recognition) [6] 50 backbone. The model was trained on the PubLaynet dataset. The evaluation showed a Mean Average Precision (mAP) of 93.6%.

Once we can detect the layout of a single page and hence detect the presence of a table, we will then proceed to the detection of the text located in this table. For text detection, there are 97k training images and 500 validation images. The training images consist of 68K real scene images and 29K synthetic images. The real scene images are collected from Baidu image search and public datasets which include: LSVT, RCTW-17, MTWI 2018, CASIA-10K, SROIE, MLT 2019, BDI, MSRATD500 and CCPD 2019. The synthetic images mainly focus on the scenarios for long texts, multi-direction texts and texts in table. The validation images are all from real scenes.

---

[11] https://developer.ibm.com/exchanges/data/all/publaynet/.

PP-OCR [21] is used to build the text detection model and they are optimized with the following optimization strategies. The evaluation of the system for text detection gave the F1-score of 85.4%.

For text recognition, there are 17.9M training images and 18.7K validation images. Among the training images, 1.9M images are real scene images, which come from some public datasets and Baidu image search. The public datasets used include LSVT, RCTW-17, MTWI 2018 and CCPD 2019. The remaining 16M synthetic images mainly focus on scenarios for different backgrounds, rotation, perspective transformation, noising, vertical text, etc. The corpus of synthetic images comes from the real scene images. All the validation images also come from the real scenes. After training on these datasets, the evaluation gave 79.4% of accuracy.

Once the text is detected and recognized, the text reconstruction allowed us to restore the text as in the table of the scientific paper. The way this works is, we draw horizontal and vertical lines based on the positions of the text in the table and their different IoU scores. The IoU is a number from 0 to 1 that specifies the amount of overlap between the predicted and ground truth bounding box.

The last step to learn Food Composition Knowledge from scientific papers is to reconstruct the tables using the text extracted. This consists of grouping the elements found in the table to obtain a new table in a tabular format (e.g., CSV). To this end, we define a naming convention of the input files and the output files:

- Naming the input source: $KNSR\_i$: this is the ith knowledge source from which tables are extracted. For instance, $KNSR\_1$ is used to designate the first PDF file from which the table is going to be extracted.
- Naming the output: each file obtained after the extraction process is in a tabular form such as CSV. Each file output is named using the ID of the input file $KNSR\_i$ plus a number j denoting the order of the apparition of the table in the file. For instance, $KNSR123\_5$ denotes the fifth table in the 123th knowledge source.

4   Knowledge Validation

The knowledge validation involved expert validation, alignment to existing biomedical ontologies and food ontologies and the matching to Wikidata and FoodOn. Globally, the evaluation showed that the dataset can be used as a relevant source to enrich food ontologies.

A Professor in Food Science and Nutrition was invited to validate a set of 100 tables selected randomly in the dataset. He was asked to check first if the terms extracted are relevant to the domain of nutrition and if the food components extracted are the ones that his research team is generally working on. He found that some tables were relevant to food and food composition. However, some tables were not relevant to food composition, and in some relevant tables, some elements (e.g. std for standard deviation) were not relevant to the food

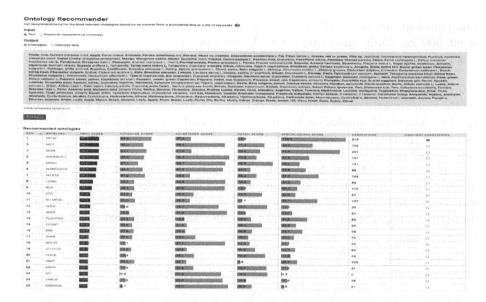

Fig. 2. Recommendation of biomedical ontologies given a set of terms extracted

component knowledge. From the tables relevant to food composition, he found that some entities were misspelled. We realized that this problem arrived during the automatic extraction of tables. We are currently cleaning all the dataset with the help of the expert by removing all the irrelevant files and correcting all the misspelled entities.

In line with the experts' validation, we evaluated the coverage of the ontology terms obtained by taking reference on biomedical ontologies hosted on BioPortal [31]. In effect, BioPortal is a biomedical ontology repository containing more than 300 biomedical ontologies, including the following 6 food ontologies: Isotopes for food science (ISO-FOOD) [3], Food-Biomarker Ontology (FOBI) [2], FoodGroupNHNS (FGNHNS)[12], OntoFood (OF)[13], The FoodOn Food Ontology (FOODON), Food Interactions with Drugs Evidence Ontology (FIDEO) [1], and Food Matrix for Predictive Microbiology (FMPM)[14].

The Ontology Recommender module was used to find the best ontologies from biomedical texts or set of keywords furnished as input in a text area [25]. This task is done according to four criteria:

- the extent to which the ontology covers the input data;
- the acceptance of the ontology in the biomedical community;
- the level of detail of the ontology classes that cover the input data;
- the specialization of the ontology to the domain of the input data.

---

[12] https://bioportal.bioontology.org/ontologies/FGNHNS.

[13] https://bioportal.bioontology.org/ontologies/OF.

[14] https://bioportal.bioontology.org/ontologies/FMPM.

The Ontology Recommender module provides users with a wizard that takes as input 500 keywords, involving a set of terms of the food knowledge extracted and providing as output the set of relevant ontologies.

Given that the text area furnished by ontology recommender can take only 500 terms, we selected randomly 500 terms from the relevant terms identified by the domain experts. The first validation consisted of matching the terms to all the biomedical ontologies hosted on Bioportal. The Fig. 2 shows that the knowledge presented as input is relevant to the biomedical domain, with some ontologies having the specialized score of more than 90%.

Surprisingly, we found that none food ontologies were present among the most relevant biomedical ontologies recommended. Thus, we decided to compare the terms extracted using food ontologies cited above. We used the same terms used during the comparison to biomedical ontologies and we obtained the Fig. 3.

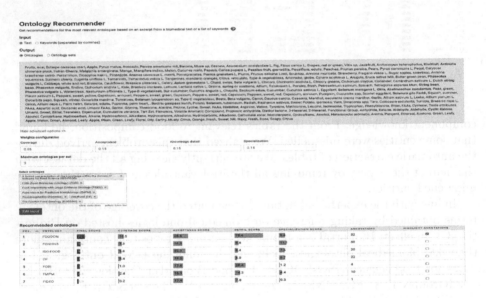

Fig. 3. Recommendation of food ontologies given a set of terms extracted

The comparison of the terms extracted to existing biomedical ontologies showed that the terms extracted were relevant to the biomedical domain in general. However, the Fig. 3 showed that food ontologies cover less terms than biomedical ontologies. Thus, we decided to match some terms extracted to the FoodOn food ontology and Wikidata Knowledge Graph.

4.1 Matching to Existing Vocabularies

cea_foodon

file	col	row	URI
KNSR3611_1	0	0	http://purl.obolibrary.org/obo/FOODON_03304644
KNSR3611_1	0	1	http://purl.obolibrary.org/obo/NCBITaxon_115466
KNSR3611_1	0	2	http://purl.obolibrary.org/obo/NCBITaxon_3750
KNSR3611_1	0	3	http://purl.obolibrary.org/obo/FOODON_03305236
KNSR3611_1	0	4	http://purl.obolibrary.org/obo/NCBITaxon_89151
KNSR3611_1	0	5	http://purl.obolibrary.org/obo/FOODON_03301357
KNSR3611_1	0	6	http://purl.obolibrary.org/obo/NCBITaxon_3494
KNSR3611_1	0	7	http://purl.obolibrary.org/obo/NCBITaxon_29760
KNSR3611_1	0	8	http://purl.obolibrary.org/obo/NCBITaxon_3489
KNSR3611_1	0	9	NIL
KNSR3611_1	0	10	http://purl.obolibrary.org/obo/NCBITaxon_3625
KNSR3611_1	0	11	http://purl.obolibrary.org/obo/NCBITaxon_151847
KNSR3611_1	0	12	http://purl.obolibrary.org/obo/NCBITaxon_29780
KNSR3611_1	0	13	http://purl.obolibrary.org/obo/FOODON_00001153
KNSR3611_1	0	14	http://purl.obolibrary.org/obo/NCBITaxon_3649
KNSR3611_1	0	15	http://purl.obolibrary.org/obo/NCBITaxon_78168
KNSR3611_1	0	16	http://purl.obolibrary.org/obo/FOODON_03415052
KNSR3611_1	0	17	http://purl.obolibrary.org/obo/FOODON_03411344
KNSR3611_1	0	18	http://purl.obolibrary.org/obo/NCBITaxon_480971
KNSR3611_1	0	19	http://purl.obolibrary.org/obo/FOODON_03412766
KNSR3611_1	0	20	http://purl.obolibrary.org/obo/NCBITaxon_4615
KNSR3611_1	0	21	http://purl.obolibrary.org/obo/NCBITaxon_22663
KNSR3611_1	0	22	http://purl.obolibrary.org/obo/NCBITaxon_88123
KNSR3611_1	0	23	http://purl.obolibrary.org/obo/NCBITaxon_13337
KNSR3611_1	0	24	http://purl.obolibrary.org/obo/FOODON_03412948
KNSR3611_1	0	25	http://purl.obolibrary.org/obo/FOODON_301693
KNSR3611_1	0	26	http://purl.obolibrary.org/obo/FOODON_119951
KNSR3611_1	0	27	http://purl.obolibrary.org/obo/NCBITaxon_58860
KNSR3611_1	0	28	http://purl.obolibrary.org/obo/FOODON_00003555
KNSR3611_1	2	0	NIL
KNSR3611_1	2	1	http://purl.obolibrary.org/obo/FOODON_03411466
KNSR3611_1	2	2	http://purl.obolibrary.org/obo/NCBITaxon_29727
KNSR3611_1	2	3	http://purl.obolibrary.org/obo/FOODON_03305806
KNSR3611_1	2	4	http://purl.obolibrary.org/obo/FOODON_03310228
KNSR3611_1	2	5	http://purl.obolibrary.org/obo/FOODON_00003410
KNSR3611_1	2	6	http://purl.obolibrary.org/obo/NCBITaxon_4045
KNSR3611_1	2	7	http://purl.obolibrary.org/obo/FOODON_03411175
KNSR3611_1	2	8	http://purl.obolibrary.org/obo/FOODON_03411314
KNSR3611_1	2	9	NIL
KNSR3611_1	2	10	http://purl.obolibrary.org/obo/NCBITaxon_13427
KNSR3611_1	2	11	http://purl.obolibrary.org/obo/NCBITaxon_4047
KNSR3611_1	2	12	http://purl.obolibrary.org/obo/FOODON_03411555
KNSR3611_1	2	13	http://purl.obolibrary.org/obo/NCBITaxon_114290
KNSR3611_1	2	14	http://purl.obolibrary.org/obo/FOODON_03411281
KNSR3611_1	2	15	http://purl.obolibrary.org/obo/NCBITaxon_4236
KNSR3611_1	2	16	http://purl.obolibrary.org/obo/FOODON_00003172
KNSR3611_1	2	17	http://purl.obolibrary.org/obo/nhv/FOODON_03000230
KNSR3611_1	2	18	http://purl.obolibrary.org/obo/FOODON_03301716
KNSR3611_1	2	19	http://purl.obolibrary.org/obo/FOODON_03305806
KNSR3611_1	2	20	http://purl.obolibrary.org/obo/NCBITaxon_65948
KNSR3611_1	4	0	NIL
KNSR3611_1	4	1	NIL
KNSR3611_1	4	2	http://purl.obolibrary.org/obo/NCBITaxon_184140
KNSR3611_1	4	3	http://purl.obolibrary.org/obo/FOODON_03411404
KNSR3611_1	4	4	http://purl.obolibrary.org/obo/FOODON_03411458
KNSR3611_1	4	5	http://purl.obolibrary.org/obo/NCBITaxon_455045
KNSR3611_1	4	6	http://purl.obolibrary.org/obo/FOODON_03317199
KNSR3611_1	4	7	http://purl.obolibrary.org/obo/FOODON_00003548
KNSR3611_1	4	8	http://purl.obolibrary.org/obo/FOODON_00003547
KNSR3611_1	4	9	http://purl.obolibrary.org/obo/FOODON_00003485
KNSR3611_1	4	10	http://purl.obolibrary.org/obo/FOODON_00003485
KNSR3611_1	4	11	http://purl.obolibrary.org/obo/FOODON_00003486
KNSR3611_1	4	12	http://purl.obolibrary.org/obo/NCBITaxon_205524
KNSR3611_1	4	13	http://purl.obolibrary.org/obo/FOODON_03411462
KNSR3611_1	4	14	http://purl.obolibrary.org/obo/FOODON_03411189
KNSR3611_1	4	15	http://purl.obolibrary.org/obo/FOODON_03000227
KNSR3611_1	6	0	NIL
KNSR3611_1	6	1	http://purl.obolibrary.org/obo/NCBITaxon_161934
KNSR3611_1	6	2	http://purl.obolibrary.org/obo/FOODON_03411227
KNSR3611_1	6	3	http://purl.obolibrary.org/obo/FOODON_03411227
KNSR3611_1	6	4	http://purl.obolibrary.org/obo/NCBITaxon_4682
KNSR3611_1	6	5	http://purl.obolibrary.org/obo/FOODON_03311757
KNSR3611_1	6	6	http://purl.obolibrary.org/obo/FOODON_03311276
KNSR3611_1	6	7	http://purl.obolibrary.org/obo/FOODON_03540568
KNSR3611_1	6	8	http://purl.obolibrary.org/obo/FOODON_00003141
KNSR3611_1	6	9	http://purl.obolibrary.org/obo/NCBITaxon_154467
KNSR3611_1	6	10	http://purl.obolibrary.org/obo/FOODON_03301449
KNSR3611_1	6	11	http://purl.obolibrary.org/obo/FOODON_00003431
KNSR3611_1	6	12	http://purl.obolibrary.org/obo/NCBITaxon_4120
KNSR3611_1	6	13	http://purl.obolibrary.org/obo/FOODON_03414254
KNSR3611_1	6	14	http://purl.obolibrary.org/obo/NCBITaxon_4460
KNSR3611_1	6	15	http://purl.obolibrary.org/obo/NCBITaxon_3711

Fig. 4. Column Entity Annotation using FoodOn

Two files were selected and annotated using Wikidata and FoodOn. To match the tables extracted to existing vocabularies, we considered the three annotations tasks described by Jimnez et al. [12]:

- Column Entity Annotation (CEA): this is the annotation of each cell of the file using the Wikidata and FoodOn entities. The Figs. 4 and 5 present the files annotated by considering the CEA task.
- Column Type Annotation (CTA): this is the annotation of the columns of the table using Wikidata and FoodOn entities. The Figs. 6 and 7 present the files annotated by considering the CTA task.
- Column Property Annotation (CPA): this is the annotation of the relation between two columns in the table using Wikidata and FoodOn properties. The Fig. 8 presents the file annotated by considering the CPA task.

These annotations showed that many elements in the files are not present in FoodOn and Wikidata. On the other hand, we found that Wikidata contains more food knowledge than FoodOn. Thus, the knowledge extracted can be a relevant source of knowledge for enriching Wikidata and FoodOn with new entities and properties.

cea_wikidata

file	col	row	URI
KNSR3611_1	0	0	http://www.wikidata.org/entity/Q1364
KNSR3611_1	0	1	http://www.wikidata.org/entity/Q33943
KNSR3611_1	0	2	http://www.wikidata.org/entity/Q18674606
KNSR3611_1	0	3	http://www.wikidata.org/entity/Q37153
KNSR3611_1	0	4	http://www.wikidata.org/entity/Q39372414
KNSR3611_1	0	5	http://www.wikidata.org/entity/Q7885904
KNSR3611_1	0	6	http://www.wikidata.org/entity/Q36146
KNSR3611_1	0	7	http://www.wikidata.org/entity/Q30046
KNSR3611_1	0	8	http://www.wikidata.org/entity/Q45757
KNSR3611_1	0	9	NIL
KNSR3611_1	0	10	http://www.wikidata.org/entity/Q163522
KNSR3611_1	0	11	http://www.wikidata.org/entity/Q15362022
KNSR3611_1	0	12	http://www.wikidata.org/entity/Q3919027
KNSR3611_1	0	13	http://www.wikidata.org/entity/Q81602
KNSR3611_1	0	14	http://www.wikidata.org/entity/Q34887
KNSR3611_1	0	15	http://www.wikidata.org/entity/Q156790
KNSR3611_1	0	16	http://www.wikidata.org/entity/Q2702783
KNSR3611_1	0	17	http://www.wikidata.org/entity/Q146281
KNSR3611_1	0	18	http://www.wikidata.org/entity/Q3411901
KNSR3611_1	0	19	http://www.wikidata.org/entity/Q158482
KNSR3611_1	0	20	http://www.wikidata.org/entity/Q1493
KNSR3611_1	0	21	http://www.wikidata.org/entity/Q13188
KNSR3611_1	0	22	http://www.wikidata.org/entity/Q1250033
KNSR3611_1	0	23	http://www.wikidata.org/entity/Q1709343
KNSR3611_1	0	24	http://www.wikidata.org/entity/Q146684
KNSR3611_1	0	25	http://www.wikidata.org/entity/Q275620
KNSR3611_1	0	26	http://www.wikidata.org/entity/Q306504
KNSR3611_1	0	27	http://www.wikidata.org/entity/Q80235
KNSR3611_1	0	28	http://www.wikidata.org/entity/Q125337
KNSR3611_1	2	0	NIL
KNSR3611_1	2	1	http://www.wikidata.org/entity/Q134609
KNSR3611_1	2	2	http://www.wikidata.org/entity/Q25694275
KNSR3611_1	2	3	http://www.wikidata.org/entity/Q2987371
KNSR3611_1	2	4	http://www.wikidata.org/entity/Q35051
KNSR3611_1	2	5	http://www.wikidata.org/entity/Q23900272
KNSR3611_1	2	6	http://www.wikidata.org/entity/Q28298
KNSR3611_1	2	7	http://www.wikidata.org/entity/Q157954
KNSR3611_1	2	8	http://www.wikidata.org/entity/Q178547
KNSR3611_1	2	9	NIL
KNSR3611_1	2	10	http://www.wikidata.org/entity/Q2544599
KNSR3611_1	2	11	http://www.wikidata.org/entity/Q41611
KNSR3611_1	2	12	http://www.wikidata.org/entity/Q42339
KNSR3611_1	2	13	http://www.wikidata.org/entity/Q178547
KNSR3611_1	2	14	http://www.wikidata.org/entity/Q45989
KNSR3611_1	2	15	http://www.wikidata.org/entity/Q36599551
KNSR3611_1	2	16	http://www.wikidata.org/entity/Q27938
KNSR3611_1	2	17	http://www.wikidata.org/entity/Q25284
KNSR3611_1	2	18	http://www.wikidata.org/entity/Q36814998
KNSR3611_1	2	19	http://www.wikidata.org/entity/Q2987371
KNSR3611_1	2	20	http://www.wikidata.org/entity/Q150452
KNSR3611_1	4	0	NIL
KNSR3611_1	4	1	http://www.wikidata.org/entity/Q2349096
KNSR3611_1	4	2	http://www.wikidata.org/entity/Q319611
KNSR3611_1	4	3	http://www.wikidata.org/entity/Q23425
KNSR3611_1	4	4	http://www.wikidata.org/entity/Q7540
KNSR3611_1	4	5	http://www.wikidata.org/entity/Q80531
KNSR3611_1	4	6	http://www.wikidata.org/entity/Q3377275
KNSR3611_1	4	7	http://www.wikidata.org/entity/Q1548030
KNSR3611_1	4	8	http://www.wikidata.org/entity/Q1548030
KNSR3611_1	4	9	http://www.wikidata.org/entity/Q1548030
KNSR3611_1	4	10	http://www.wikidata.org/entity/Q1548030
KNSR3611_1	4	11	http://www.wikidata.org/entity/Q165308
KNSR3611_1	4	12	http://www.wikidata.org/entity/Q2670836
KNSR3611_1	4	13	http://www.wikidata.org/entity/Q7533
KNSR3611_1	4	14	http://www.wikidata.org/entity/Q161180
KNSR3611_1	4	15	http://www.wikidata.org/entity/Q23501
KNSR3611_1	6	0	NIL
KNSR3611_1	6	1	http://www.wikidata.org/entity/Q165191
KNSR3611_1	6	2	http://www.wikidata.org/entity/Q30056
KNSR3611_1	6	3	http://www.wikidata.org/entity/Q83124
KNSR3611_1	6	4	http://www.wikidata.org/entity/Q23400
KNSR3611_1	6	5	http://www.wikidata.org/entity/Q1807269
KNSR3611_1	6	6	http://www.wikidata.org/entity/Q23485
KNSR3611_1	6	7	http://www.wikidata.org/entity/Q311356
KNSR3611_1	6	8	http://www.wikidata.org/entity/Q743144
KNSR3611_1	6	9	http://www.wikidata.org/entity/Q790172
KNSR3611_1	6	10	http://www.wikidata.org/entity/Q10998
KNSR3611_1	6	11	http://www.wikidata.org/entity/Q7224565
KNSR3611_1	6	12	http://www.wikidata.org/entity/Q37937
KNSR3611_1	6	13	http://www.wikidata.org/entity/Q71549
KNSR3611_1	6	14	http://www.wikidata.org/entity/Q227997
KNSR3611_1	6	15	http://www.wikidata.org/entity/Q3384

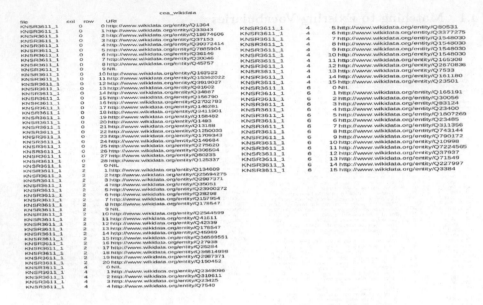

Fig. 5. Column Entity Annotation using Wikidata

cta_foodon

file	col	URI
KNSR3611_1	0	http://purl.obolibrary.org/obo/FOODON_00002141
KNSR3611_1	2	http://purl.obolibrary.org/obo/FOODON_03411036
KNSR3611_1	4	http://purl.obolibrary.org/obo/FOODON_03411036
KNSR3611_1	6	http://purl.obolibrary.org/obo/FOODON_03411036

Fig. 6. Column Type Annotation using FoodOn

cta_wikidata

file	col	URI
KNSR3611_1	0	http://www.wikidata.org/entity/Q16521
KNSR3611_1	2	http://www.wikidata.org/entity/Q11004
KNSR3611_1	4	http://www.wikidata.org/entity/Q11004
KNSR3611_1	6	http://www.wikidata.org/entity/Q11004

Fig. 7. Column Type Annotation using Wikidata

cpa_foodon

file	col0	colx	URI
KNSR3611_1	0	1	NIL
KNSR3611_1	0	2	NIL

Fig. 8. Column Property Annotation using FoodOn

5 Conclusion

This paper showed how Food Composition Knowledge can be learned from Food Composition Tables published in scientific literature. This consists of table detection, text recognition, text extraction and table reconstruction using Deep Learning techniques trained on large datasets. Once extracted, we proceed to the validation by an expert in Food Science and Nutrition. Thereafter, we compared some terms extracted to existing biomedical ontologies and existing food ontologies using the ontology recommender module of BioPortal. The matching with existing biomedical and food ontologies show that these knowledge can be used to enrich the latter.

Currently, the Food Composition Knowledge extracted is being used to build a Food Composition Knowledge Graph that we named TSOTSAGraph. The purpose of this graph is to recommend food to people given their health profile.

Acknowledgements. We are grateful to neuralearn.ai for having provided the video tutorial and all the machine learning source code needed to extract tables from scientific papers. We are also grateful to Kangsi Germain, Professor in Food Science and Nutrition for the contribution to the validation of information extracted.

References

1. Bordea, G., Nikiema, J.N., Griffier, R., Hamon, T.: FIDEO: food interactions with drugs evidence ontology. In: ICBO/ODLS (2020)
2. Castellano-Escuder, P., González-Domínguez, R., Wishart, D.S., Andrés-Lacueva, C., Sánchez-Pla, A.: FOBI: an ontology to represent food intake data and associate it with metabolomic data. Database J. Biol. Databases Curation **2020** (2020)
3. Eftimov, T., Ispirova, G., Potovcnik, D., Ogrinc, N., Seljak, B.: ISO FOOD ontology: a formal representation of the knowledge within the domain of isotopes for food science. Food Chem. **277**, 382–390 (2019)
4. International Network of Food Data Systems (INFOODS): Sustainable development goals (2022). https://www.fao.org/infoods/infoods/en/
5. Greenfield, H., Southgate, D.A.: Food composition data: production, management, and use. Food & Agriculture Organization (2003). https://doi.org/10.1007/978-1-4615-3544-7
6. He, K., Zhang, X., Ren, S., Sun, J.: Deep residual learning for image recognition. CoRR abs/1512.03385 (2015). http://arxiv.org/abs/1512.03385

7. He, M., et al.: ICPR 2018 contest on robust reading for multi-type web images. In: 2018 24th International Conference on Pattern Recognition (ICPR), pp. 7–12 (2018). https://doi.org/10.1109/ICPR.2018.8546143
8. He, W., Zhang, X.Y., Yin, F., Liu, C.L.: Multi-oriented and multi-lingual scene text detection with direct regression. IEEE Trans. Image Process. (TIP) **27**(11), 5406–5419 (2018)
9. Huang, X., et al.: PP-YOLOv2: a practical object detector. CoRR abs/2104.10419 (2021). https://arxiv.org/abs/2104.10419
10. Huang, Z., et al.: ICDAR 2019 competition on scanned receipt OCR and information extraction. In: 2019 International Conference on Document Analysis and Recognition (ICDAR), pp. 1516–1520 (2019). https://doi.org/10.1109/ICDAR.2019.00244
11. Informatics, D.F.: Langual - the international framework for food description (2020). https://www.langual.org/Default.asp
12. Jiménez-Ruiz, E., Hassanzadeh, O., Efthymiou, V., Chen, J., Srinivas, K.: SemTab 2019: resources to benchmark tabular data to knowledge graph matching systems. In: Harth, A., et al. (eds.) ESWC 2020. LNCS, vol. 12123, pp. 514–530. Springer, Cham (2020). https://doi.org/10.1007/978-3-030-49461-2_30
13. Jiomekong, A.: Comparison of food composition tables/databases (2022). https://orkg.org/comparison/R206121/
14. Jiomekong, A.: Food information engineering: a systematic literature review. Proc. AAAI Conf. Artif. Intell. **37**(13), 15441 (2023). https://doi.org/10.1609/aaai.v37i13.26808. https://ojs.aaai.org/index.php/AAAI/article/view/26808
15. Jiomekong, A., Camara, G., Tchuente, M.: Extracting ontological knowledge from Java source code using hidden Markov models. Open Comput. Sci. **9**(1), 181–199 (2019)
16. Jiomekong, A., et al.: A large scale corpus of food composition tables. In: SemTab@ISWC (2022)
17. Jiomekong, A., Uriel, M., Tapamo, H., Camara, G.: Semantic annotation of TSOT-SAtable dataset. In: SemTab@ISWC (2023)
18. Karatzas, D., Mestre, S.R., Mas, J., Nourbakhsh, F., Roy, P.P.: ICDAR 2011 robust reading competition - challenge 1: reading text in born-digital images (web and email). In: 2011 International Conference on Document Analysis and Recognition, pp. 1485–1490 (2011). https://doi.org/10.1109/ICDAR.2011.295
19. Khalis, M., et al.: Update of the Moroccan food composition tables: towards a more reliable tool for nutrition research. J. Food Compos. Anal. **87**, 103,397 (2020)
20. Kouebou, C., et al.: A review of composition studies of Cameroon traditional dishes: macronutrients and minerals. Food Chem. **140**(3), 483–494 (2013)
21. Li, C., et al.: PP-OCRv3: more attempts for the improvement of ultra lightweight OCR system (2022)
22. Makwana, Y., Iyer, S.S., Tiwari, S.: The food recognition and nutrition assessment from images using artificial intelligence: a survey. ECS Trans. **107**(1), 3547 (2022)
23. Nayef, N., et al.: ICDAR 2019 robust reading challenge on multi-lingual scene text detection and recognition - RRC-MLT-2019. In: 2019 International Conference on Document Analysis and Recognition (ICDAR), pp. 1582–1587 (2019). https://doi.org/10.1109/ICDAR.2019.00254
24. O'Gorman, L.: The document spectrum for page layout analysis. IEEE Trans. Pattern Anal. Mach. Intell. **15**(11), 1162–1173 (1993). https://doi.org/10.1109/34.244677

25. Romero, M.M., Jonquet, C., O'Connor, M.J., Graybeal, J., Pazos, A., Musen, M.A.: NCBO ontology recommender 2.0: an enhanced approach for biomedical ontology recommendation. J. Biomed. Semant. **8**(1), 21:1–21:22 (2017). https://doi.org/10.1186/s13326-017-0128-y

26. Shi, B., et al.: ICDAR 2017 competition on reading Chinese text in the wild (RCTW-17). In: 2017 14th IAPR International Conference on Document Analysis and Recognition (ICDAR), vol. 01, pp. 1429–1434 (2017). https://doi.org/10.1109/ICDAR.2017.233

27. Sun, Y., Liu, J., Liu, W., Han, J., Ding, E., Liu, J.: Chinese street view text: large-scale Chinese text reading with partially supervised learning. In: 2019 IEEE/CVF International Conference on Computer Vision (ICCV), pp. 9085–9094. IEEE Computer Society, Los Alamitos, CA, USA (2019). https://doi.org/10.1109/ICCV.2019.00918. https://doi.ieeecomputersociety.org/10.1109/ICCV.2019.00918

28. UN: Sustainable development goals (2015). https://sdgs.un.org/goals

29. Usip, P.U., Udo, A., Ijebu, F.F., Tiwari, S.: A review on multilingual food recommendation systems for critical medical conditions in pregnancy care (2022)

30. Watanabe, T.: Food composition tables of japan and the nutrient table/database. J. Nutr. Sci. Vitaminol. **61**(Supplement), S25–S27 (2015)

31. Whetzel, P.L., et al.: BioPortal: enhanced functionality via new web services from the national center for biomedical ontology to access and use ontologies in software applications. Nucleic Acids Res. **39**, 541–545 (2011). https://doi.org/10.1093/nar/gkr469

32. Xu, Z., et al.: Towards end-to-end license plate detection and recognition: a large dataset and baseline. In: Ferrari, V., Hebert, M., Sminchisescu, C., Weiss, Y. (eds.) ECCV 2018. LNCS, vol. 11217, pp. 261–277. Springer, Cham (2018). https://doi.org/10.1007/978-3-030-01261-8_16

33. Yadav, S., Powers, M., Vakaj, E., Tiwari, S., Ortiz-Rodriguez, F., Martinez-Rodriguez, J.L.: Semantic based carbon footprint of food supply chain management. In: Proceedings of the 24th Annual International Conference on Digital Government Research, pp. 657–659 (2023)

34. Yao, C., Bai, X., Liu, W., Ma, Y., Tu, Z.: Detecting texts of arbitrary orientations in natural images. In: 2012 IEEE Conference on Computer Vision and Pattern Recognition, pp. 1083–1090. IEEE (2012)

35. Zhong, X., Tang, J., Yepes, A.J.: PubLayNet: largest dataset ever for document layout analysis. In: 2019 International Conference on Document Analysis and Recognition (ICDAR), pp. 1015–1022. IEEE (2019). https://doi.org/10.1109/ICDAR.2019.00166

Design and Analysis of an Algorithm Based on Biometric Block Chain for Efficient data sharing in VANET

Arpit Namdev[✉] and Harsh Lohiya

Sri Satya Sai University of Technology, Sehore, Madhya Pradesh, India
namdev.arpit@gmail.com

Abstract. VANETs allow cars to interact with each other and infrastructure, improving road safety and traffic efficiency. However, safe and effective vehicle-road infrastructure data exchange is a major obstacle to VANET implementation. VANET data exchange methods are vulnerable to cyberattacks and lack identity verification. This study uses biometrics and blockchain to solve VANET data sharing problems. The system creates a unique and tamper-resistant identity using biometric data from the vehicle's driver or authorized workers. A blockchain-based distributed ledger anchors this identification, assuring data transaction immutability and transparency. Algorithm creation and analysis need numerous phases. First, a secure biometric recognition system is constructed. Next, a permissioned blockchain framework creates a decentralized network for allowed data sharing. Consensus techniques preserve data integrity and protect sensitive information in the proposed blockchain. Extensive simulations and real-world trials assess the algorithm's efficiency and efficacy. Data sharing speed, transaction throughput, and security are compared to typical VANET data sharing methods. The biometric blockchain algorithm excels in data privacy, authentication speed, and cyber-attack resistance. This study has major implications for real-world VANET implementation. Biometric authentication and blockchain technologies improve data security and enable reliable and efficient vehicle-infrastructure connections. This study helps intelligent transportation systems become safer and smarter by addressing data integrity and identity verification issues.

Keywords: Vehicular Ad hoc -Networks (VANET) · Biometric · Blockchain · Internet

1 Introduction

Today's vehicle network management relies heavily on communication technologies. In the past, vehicles' many internal subsystems were all managed by a single unified system architecture, necessitating the development of wired communication protocols to ensure seamless operation. As a effect of being hardwired, the cost of the automobile's scheme and upkeep goes up. In order to maximise efficiency, today's automotive networks rely on wireless protocols to transmit and receive data between vehicles and the outside world.

© The Author(s), under exclusive license to Springer Nature Switzerland AG 2023
S. Tiwari et al. (Eds.): AI4S 2023, CCIS 1907, pp. 104–118, 2023.
https://doi.org/10.1007/978-3-031-47997-7_8

Intelligent transportation systems [1–3], carpooling [4] and the use of fifth-generation small-cell schemes [5, 6] are just a few examples of the many applications for VANET, a new efficient wireless technology. A large number of wireless sensors, converter devices, and mapping units make up each vehicle in a VANET system. An OBU (On-Board Unit) and an RSU (Road Side Unit) are the two interface modules that make up the VANET system (RSU). The OBU module is installed within the car and acts as a hub for all wireless sensors. Each roadside structure has a transmitter and a receiver installed so that the OBU module in each car can communicate with the other cars on the road. When a car connects to a VANET network, the numerous sensors installed in it collect data specific to that car and relay that data to the RSU section. When the automobile's OBU segment and RSU module communicate effectively with one another, accidents are avoided. Both the OBU and the RSU will be compromised by the presence of unauthorised individuals. There must be some sort of protection system in place between the OBU and the RSU components. Using VANET systems, in which vehicles share data with their neighbors, greatly reduces the number of accidents that occur. The VV component of the VANET scheme is responsible for the dissemination of evidence from vehicle to vehicle, and the VTI module is responsible for the distribution of information from vehicles to infrastructure. The VANET system allows for data to be transmitted from one vehicle to a controller or centralized system using the VI module. Due to the active landscape of the real scenario in which the VANET operates, the topology system is constantly evolving to accommodate new vehicle distances and locations. For instance, ambient noise can distort the transmission of data between vehicles [7–9]. Rugged VANET environments, such as those found in vehicles, are susceptible to outside attacks such as eavesdropping and data hacking. A vehicular ad hoc network (VANET) wherein every vehicle communicates wirelessly every device available in network. In a VANET, an attacker targets both the vehicles and the controller in a coordinated effort. The RSU module is a principal controller in VANET systems that link multiple vehicles to a single network. Most of the damage done by the attackers was to the communication between vehicles and the centralized controller.

The remainder of the paper is organized as follows In Sect. 2, we provide an overview of related work in the literature. In Sect. 3, we provide our proposed work base on block chain technology In Sect. 4, we give the numerical results of a detailed performance of the algorithm under different scenarios In Sect. 5, we state some conclusions and give some commentary on the future.

2   Literature Review

Extensive research has been conducted in recent years on the topic of managing communications between vehicles and ensuring the highest quality of service in a vehicular ad hoc network. To address these issues, Guoqiang Mao et al. (2018) proposed a novel passive multi-hop clustering method. (PMC). The PMC algorithm is based on the concept of a multi hop clustering algorithm, which ensures that all clusters are consistently covered. In this article, we present a priority-based neighbour following method for selecting the most suitable neighbour nodes to invite into the same cluster as the cluster head. Using this method, you can rest assured that the connections between clusters will be robust and secure. By ensuring node stability and then appointing the most stable node within the N-hop range as the cluster leader, we can greatly improve the stability of the clustering process. To further improve the cluster's dependability and resilience, the collection fusion process is executed during the collection maintenance phase. They run extensive tests in the NS2 environment comparing the PMC approach to the N-HOP, VMaSC, and DMCNF processes to verify the PMC approach's efficacy [1].

A. Lawal et al. (2015) In this research, the author proposed an innovative approach to multi-constrained QoS steering in vehicular networks by employing a bunching strategy. The procedure's plan makes use of a number of quality-of-service (QoS) measurements in addition to stability metrics in order to discover and create a stable path to the final destination. To do so, we first determine the necessary QoS provision values before deciding on a safe, dependable, and efficient route to take. The proposed system was evaluated using the NCTUns 6.0 system trainer. According to the findings, the planned strategy dramatically reduces the number of failed connections, the amount of overhead caused by routing, and the total amount of time it takes for a note to travel its endpoint. The main goal of this work is to investigate the challenges of providing packet-level QoS for real-time traffic in vehicle networks and to ensure the reliability of routing safety traffic data. They planned to get around the problems by spreading the priority queues out across different channels. Due to the unpredictability of high-mobility vehicles, both the Control Channel and Service Channel DSRC were designed with seven-channel bands and their routing path was optimised to reduce latency and connection time in QoS caused by packet retransmission [2].

Mirzaee et al. (2018) proposed a voting strategy that would improve the reliability of message voting. Voting time and network data overhead are both reduced. The planned process is "spoofed" in the software design verbal NS2. When compared to other similar approaches in the prose, the experimental consequences of this education display that the proposed techniques improve decision accuracy by a range of 6% to 30% depending on factors such as circulation volume, amount of bumps, and period of action times. It decreases the amount of sachets sent from 1 percent to 9 percent, depending on the conditions. To guarantee message security, instant node connectivity, and quick data transmission, they attempted to manage clustering via multi-channel communications. The optimal cluster size, the number of cluster heads, the nodes' average speed, and their orientation all play a role in determining the clustering parameters. The resulting clusters are optimal for use in large-scale environments [3].

Badia Bouhdid, et al. (2015) proposed research which offerings a unique crowding approach for vehicular ad hoc systems by considering not only node residence but also node movement. The goal of the planned scheme is to construct stable clusters by falling the expense of re-clustering, increasing the lifespan of existing clusters, and decreasing the average distance between CHs and other members of the cluster. Importantly, this approach is applicable to both single and multiple CHs. In comparison to three other popular methods, the clustering approach performs better in simulations. If the normal reserve amid CHs and the members of their cluster is decreased, collisions may occur. Their help ensures that the overall average distance stays reasonable. To make clusters in VANETs last longer, they introduced a beacon-based clustering technique. They decided to reorganise the clusters according to a new aggregate local mobility criterion [4].

Hafid et al (2016) proposed a novel work which is based on their movability, vehicles are clustered into non-overlapping groups of varying sizes using DHCV, a D-hop clustering method introduced in this article. D-hop clustering builds groups so that no vehicle is more than D hops from the group's leader. Each vehicle in a multi-hop cluster will select a leader based on how mobile it is expected to be relative to its D-hop nationals. The procedure can be repeated at fixed pauses or whenever there is a shift in the network topology. The algorithm tends to re-elect the remaining cluster leaders whenever the web assembly shifts, which is one of its distinguishing structures. The efficacy of our clustering method has been demonstrated via extensive simulation results obtained in a wide variety of settings. [8] The authors provide a workaround for the dynamic VANET by proposing a method of electing the cluster head using a variety of parameters. Excellence of service metrics are taken into account when choosing the cluster leader. Direction, velocity, time spent at the helm of a cluster previously, cluster density, vessel transfer ratio, network longevity, node degree, and transmission range are all factors to consider [5].

Erritali et al (2017) and colleagues [10] use MOVE Tool and SUMO to simulate a Dark Hovel dose in a VANET setting, employing a realistic movement archetypal they've developed. They go on to recommend a clustering strategy for identifying and counteracting the black hole attacker node. As part of their project, they must create a mobility model that uses SUMO and the MOVE Tool to simulate continuous road traffic in order to construct a realistic simulation. To make the attack more realistic, the researchers built a Black Hole attack private the model and analysed the results to see how it affected the network's ability to send and accept data. Then, they proposed a cluster-based method for identifying the malicious node and removing it from the network based on an analysis of the algorithm's complexity [6].

Suzi Iryanti, et al. (2017) study presents a new variant of the Harmless Assembling Procedure (M-SCA), where security space is employed to extend the band phase within the cluster maintenance stage. QoS settings, substitute communication inter-arrival period, and above are simulated in NS2 to determine the efficacy of the proposed scheme. In order to reduce the likelihood of chain collisions, they propose using a clustering technique that takes advantage of the advantages of such networks and ensures that urgent safety signals are sent out quickly. Throughput, container transfer share, endwise latency, and overhead are just a few of the metrics that are measured to evaluate the proposed clustering system's efficiency. Two alternative clustering algorithms are compared with the presented method [7].

Researchers Reza Rafeh et al (2016) and coworkers proposed a distributed, scalable, and proficient routing method that makes use of clusters. The proposed method takes into account both the vehicle's speed deviation and the time left until coming when deciding which cluster head to use. Replication marks indicate that the planned method has lower Endwise latency than the CBLR procedure. It is recommended to use a CBRP-based cluster-based routing procedure. This method divides the network's nodes into a set of overlapping or distinct clusters, each with a 2-hop diameter, in a decentralised fashion. An approach to routing based on clusters was proposed [8].

Boucher Mazak, Hitchin Terumi, Mohamed Tales, Elhabib Benlahmar, et al (2015) and others. Proposed a YATES method which is used to present a model for determining the worth of stable nodes in this paper. The goal of this strategy is to break up stable clusters. Distance, probability, and the difference in speeds are used in the Yates method to calculate a node's stability value, which is then incorporated into a model. The proposed model boasts more stable clusters, as evidenced by a low rate of CH change, a long CH lifetime, and a long lifespan for cluster members. Their findings support YATES as a viable strategy for future routing system creation. Space, speed difference, and probability factors were used to estimate the stability of each node in their clustering method. The approach may correspond to the requirements of current bunching beating rules in relations of constructing constant channels connecting various swellings [9].

Together, Mohamed Talea et al (2017) and his associates they give their novelist efforts on VANET beating protocols and provides a brief overview of several of them. A thorough examination of routing protocols is provided. All treatments were put through their paces on the NS-2 as we varied the mobility and load that was supplied. Average round-trip latency and bandwidth are compared as a means of assessing performance. They establish both the strengths and weaknesses of various beating conventions in a VANET location. The researchers set out to find out which routing mechanism would work best in the fluid and dynamic setting of a VANET. In this article, the cluster-based AMACAD and MOBIC defeating conventions are compared to the more traditional responsive, practical, and location defeating conventions. Cluster-based beating conventions are equivalent to one another in positions of normal endways latency, container sending proportion, and bandwidth. The most recent research in VANET security is summarised in the table below, where we discuss how different neural network, machine learning, and block chain methodologies are used to ascertain the accuracy of models [10]

Zeng et al. (2020) proposes a Trusted Ledger Model (TLM) to ensure the consistency of multiple data resources in a low trust distributed computing environment. The paper implements the Fengyi system based on the TLM to provide authentication and encrypted communication services with the ledgers. The paper deploys the Fengyi system on three different platforms to check its effectiveness and efficiency in ensuring trusted data sharing in VANETs.he paper proves that traceability, integrity, and non-repudiation combined are sufficient conditions of consensus, which are equivalent to accountability for data sharing in VANETs.

The main contributions of the paper are as follows (Table 1):

Table 1. Shows the comparative study of vehicular ad-hoc network.

Ref. No	Author's name	Publication detail	Advantages	Tools
[27]	Kandli Et al	ITS: An IEEE Transactions Journal, 2021	An Improved K-Means Clustering Algorithm and a Continuous Hopfield Network Form a Novel Hybrid Routing Protocol for VANETs	NS2 (release 2.35) network simulator
[28]	Alharthi el al	IEEE, 2021	biometrics lump sequence (BBC) background for safe data exchange between vehicles in VANET and reliable long-term storage of vital records. Novelist uses biometric information to verify the sender's identity and protect the sender's privacy within the proposed framework	Omtet++,Vinus,SUMO
[29]	Hossain Et al	IEEE transaction 2021	Within the scope of this paper, we propose a Multi-Objective Harris Hawks Optimization (2HMO-HHO) based 2-Hop routing algorithm that chooses optimal forwarders between the starting and ending points of a trip	OMnet++ and SUMO
[30]	Kazi et al.	IEEE 2021	Reliable Group of Vehicles (RGoV) in VANET	NS2 and MOVE
[31, 32, 33]	Ikram Uddin et al. Temurnikar et al.	IEEE, 2020	The IoT uses a Left-Right-Front Caching Strategy for Automotive Networks	NS3 and SUMO
[32, 34, 35, 36, 37, 38]	Funderburg et al. Temurnikar et al.	IEEE 2021	The work presented here introduces a practical, pairing-free signature scheme for VANETs that protects against identity forgery even in the face of insider attacks and does not require the use of a physical device to detect tampering	-----

3 Proposed Work

Vehicle Deployment and Network Construction Using TA, RSU, Block Chain, and MVD (Motor Vehicle Department) The lump chain-based VANET safe and trustworthy records distribution organisation mechanism is described in this part [34, 35].

Vehicles: Vehicles are equipped with numerous beams, as well as packing and wireless email segments, and comunicate with one another through wireless networks; these modules are the primary section of the SDN records smooth. Using RSU, vehicles can send information about the current road conditions to higher authorities for evaluation and broadcast the information to other nearby vehicles. The architectural plan alludes to C-V2X, which is a cellular network-based wireless statement system for automobiles.

UAVs: Unmanned aerial vehicles (UAVs) can function as authority lumps, bringing authority figuring facilities to outlying zones. All the data processing, storage, and wireless communication modules required between the SDN controller and the vehicle are contained within the Relay Switching Units (RSUs). Stations of origin: As the backbone of the software-defined networking (SDN) control plane, base stations are equipped with mobile edge computing (MEC) capabilities and distributed across a network. There are also records cables and contract chains associated with each base station. The data chain's responsibility is to keep track of the information passed between vehicles, while the agreement cable's is to include automobile data and automobile elective. The MEC is responsible for carrying out corresponding set matching, preventing any computation activities, and providing storage resources for storing evidence on the lump cable [36, 37].

Certification authority (CA): The CA is generally trusted as an authoritative source that communicates vital information about each vehicle in order to create a personalised block chain address for each user. In the event of a spoofed message being sent by a car user during a data transfer, the CA can verify the sender's true identity.

Cloud: The cloud receives the car-to-cloud messages after they have been temporarily stored at the edge nodes. Because some outlying regions are not within the SDN control area and therefore cannot receive messages in a timely manner, the final data sharing must additionally broadcast messages to the remaining vehicles via the cloud. This could lead to vehicles being more cognizant of circulation situations and enhancing the security and efficiency of scheme communication.

To protect user privacy and withstand complicity attacks, a cloud-based block chain-based voting consensus mechanism for PSI empowers vehicles with similar characteristics to make subjective data accuracy evaluations, improving data evaluation reliability. Second, this research employs block chain technology to establish a file packing organization and an elective smart agreement to protect data from manipulation. The car sharing data will make a transaction and pack it into a block, similar to signature verification, and the transaction address may verify the vehicle's identification, improving system confidence. This article's third addition is using word2vec and cosine similarity comparison to quickly deduplicate suspected duplicate data, easing blockchain maintenance and reducing blockchain storage overhead. To address VANET communication needs, this research provides a unique C-V2X communication structure employing MEC, SDN, and other technologies. This essay concludes with a block chain award distribution procedure. The block chain administrator node sends a transfer transaction or activates a smart contract. SDN with edge computing provide efficient information sharing and network performance in a VANET (Vehicular Ad-Hoc Network) system.

SDN is a networking paradigm that separates the control plane from the data plane in network devices such as switches and routers. It centralizes the network management and control functions in a separate entity known as the "SDN controller". This controller is responsible for making intelligent decisions about how data should be forwarded across the network.

In the context of VANET, SDN can be utilized to optimize communication, manage traffic, and enhance overall network efficiency. Here's how SDN is applied in VANETs:

a. Centralized Network Control: The SDN controller can have a global view of the entire VANET, including the position and status of vehicles. This allows for better traffic management and efficient routing decisions.
b. Dynamic Routing: With real-time information on traffic conditions and road statuses, the SDN controller can dynamically update the routing paths for data transmission, ensuring that information reaches its destination via the most optimal route.
c. Quality of Service (QoS) Management: SDN enables the prioritization of certain types of data, such as emergency messages, over others, ensuring that critical information is delivered with minimal delay.
d. Network Resilience: SDN allows for quick adaptability in case of network disruptions or changes in VANET topology. The controller can rapidly reconfigure the network to maintain connectivity and data exchange.

Edge Computing: Edge computing is a distributed computing paradigm that brings computation and data storage closer to the location where it is needed, reducing latency and reliance on centralized data centers. In VANETs, edge computing can be applied to enhance information exchange and support various services:

a. Low Latency Data Processing: Edge computing nodes deployed at the roadside infrastructure or in nearby vehicles can process time-sensitive data locally. This reduces the time it takes to analyze and respond to critical events, such as collision warnings or traffic updates.
b. Data Offloading: Edge nodes can offload data processing tasks from the central cloud or data centers. This helps in reducing the backhaul traffic and conserves network bandwidth.
c. Scalability: Edge computing provides a scalable infrastructure that can cope with the increasing volume of data generated by vehicles in the VANET environment.
d. Privacy and Security: Edge computing can be used to perform data filtering and anonymization locally, enhancing privacy and security in the network.

By combining SDN and edge computing in VANET infrastructure, it's possible to create an intelligent and efficient network that can handle the dynamic and data-intensive nature of vehicular communication. The coordination between centralized SDN control and distributed edge computing nodes helps optimize traffic management, improve data processing, and provide a more reliable and responsive VANET ecosystem.

Listed below are three distinct but complementary contributions.

(1) By joining SDN, MEC, block chain, and extra tools, we present a game-changing architecture for Internet VANET networks that guarantees the security and efficiency of data transfers.
(2) In order to ensure the privacy of both users' data and their vehicles, a PSI protocol was developed that combines hash and random number and is executed at the network's edge nodes. The protocol also decreases matching overhead.
(3) Using smart contracts and consensus mechanisms, we create a block chain system for the internet-VANET and guarantee the immutability, authenticity, and photograph of all transactions and data (Fig. 1).

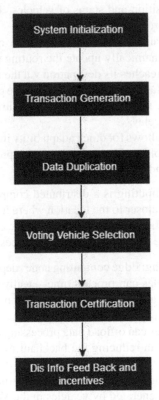

Fig. 1. Flow Diagram of Proposed Work

A. System Initialization

Among the most cutting-edge and sophisticated vehicles currently available Vi initially logs into the system to obtain a key pair (PK vi and SKvi) to verify its legal identification, which the CA in the SDN application layer may then distribute. Using the secp256k1 elliptic curve and Ethernet's ECDSA method, the public and private keys can be created as 64-bit hexadecimal strings, guaranteeing the key's authenticity (Fig. 2).

B. Transaction Generation

All financial dealings have been sent. If a automobile Vp needs to refer information Mp to extra vehicle, it need major create a transaction Tx and send the details of Tx to the receiving vehicle.

```
struct VANETTransaction {
    sender: public_key,      // Public key of the sender vehicle
    receiver: public_key,    // Public key of the receiver vehicle or infrastructure unit
    data: encrypted_data,    // Encrypted data payload
    timestamp: timestamp,    // Timestamp of the transaction
    signature: digital_signature, // Digital signature to verify authenticity
}
struct VANETBlock {
    index: integer,          // Index or height of the block in the blockchain
    previous_hash: hash,     // Hash of the previous block
    timestamp: timestamp,    // Timestamp of the block creation
    transactions: array[VANETTransaction], // Array of transactions in the block
    hash: hash,              // Hash of the current block
    nonce: integer,          // Nonce value for mining (Proof-of-Work)
}
```

```
                    initialize_blockchain():
                        genesis_block = {
                            index: 0,
                            previous_hash: "0",
                            timestamp: current_time(),
                            transactions: [],
                            hash: calculate_hash(0, "0", current_time(), [], 0),
                            nonce: 0
                        }
                        blo// Create a new VANET transaction
                    create_transaction(sender, receiver, data):
                        transaction = {
                            sender: sender.public_key,
                            receiver: receiver.public_key,
                            data: encrypt_data(data, receiver.public_key),
                            timestamp: current_time(),
                            signature: sign_data(data, sender.private_key)
                        }
                        return transactionckchain = [genesis_block]
```

C. Data Duplication

When a user uploads data, it often contains similar recent data. To avoid using unnecessary computational storage space, users should verify the data against the most recent data by timestamps before uploading. Common phrase similarity metrics include vector space cosine similarity, which uses the cosine of the angle between two vectors in vector space to quantify the degree of dissimilarity between two speakers.

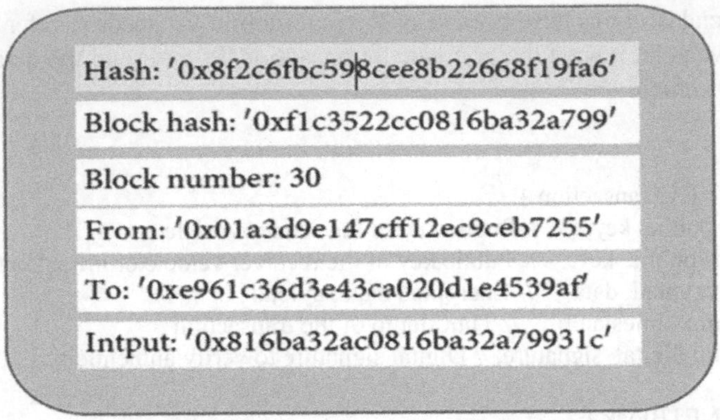

Hash: '0x8f2c6fbc598cee8b22668f19fa6'

Block hash: '0xf1c3522cc0816ba32a799'

Block number: 30

From: '0x01a3d9e147cff12ec9ceb7255'

To: '0xe961c36d3e43ca020d1e4539af'

Intput: '0x816ba32ac0816ba32a79931c'

Fig. 2. Data Transaction Information

D. Voting Vehicle Selection

After Vp's identity has been verified, he can begin collecting Vi data for matching purposes; those Vi members who have voting rights will coalesce to form Vq. Because of the high volume of vehicles participating in VANET, a cloud-assisted PSI protocol needs to be proposed for this procedure.

E. Transaction Certification

After a vehicle sends its first message, new transactions are created, but they are not yet included in blocks until they go through the consensus process and the mining operation performed by miners. The latency overhead of the system may change after blockchain technology is implemented.

F. Disinformation Feed Back and Incentives

Cars can vote for harmful fake messages as a group, which can mislead other automobiles and compromise circulation security, and communications like highway evidence may become out-of-date as time passes.

4 Experimental Result

A. Experimental Settings

The experimental PC setup includes a Windows operating system, a Core i3 processor, and 16 GB of RAM. NS2 and Jupiter Notebook, both of which are based in Python, are used as the simulation platform [38].

B. Storage Overhead

If blockchain technology is implemented in VANET, data storage utilisation rates will rise. To compare, Zeng et al. [25, 39] introduced a blockchain-based data sharing system called Fengyi, but they didn't use smart contracts, so the total amount of memory used by Fengyi for shared data is only about 1.4 kilobytes. One data sharing in Fengyi takes about 1.4 kb of memory, while announcement message sharing is designed to cut down on the time and space needed for encryption and decryption (Figs. 3 and 4).

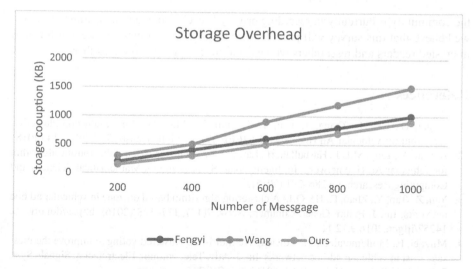

Fig. 3. Storage Overhead of Shared Data

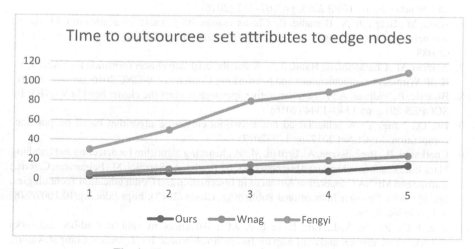

Fig. 4. Time to outsource set attribute to edge node

5   Conclusion

Scientists are increasingly concerned about how machine learning impacts security, privacy, fairness, or explain ability. A security model and a unified taxonomy of the many types of attacks were presented here based on the features of recent privacy-related attacks. In our study, we identified basic design patterns and contrasted them. State-of-the-art research at the time led to this. Up until now, experimental investigations on the variables affecting privacy leaks have offered insightful information. However, surprisingly few studies test attacks on actual data sizes and deployments. Research on security, explain ability, and justice is needed. In spite of the experimental state, the community is currently in regarding privacy breaches in machine learning systems, we believe that this survey will provide the background information essential for both interested readers and researchers who wish to investigate this issue further.

References

1. Zhang, D., Ge, H., Zhang, T., Cui, Y.-Y., Liun, X., Mao, G.: New multi-hop clustering algorithm for vehicular Ad Hoc networks. IEEE Trans. Intell. Transp. Syst. **20**, 1–14 (2018)
2. Aminu, A., Tang, M.L.J., Hasbullah, H., Lawal, I.A.: A cluster-based stable routing algorithm for vehicular Ad Hoc network. In: International Symposium on Mathematical Sciences and Computing Research, pp 286–291 (2015)
3. Yan, Z., Tao, Y., Zhao, F., He, Q.J.: A clustering algorithm based on zone in vehicular ad hoc networks. Int. J. Future Gener. Commun. Netw. **9**(12), 117–128 (2016). https://doi.org/10.14257/ijfgcn.2016.9.12.11
4. Mirzaei, F., Mollamotalebi, M.: An adaptive multi-hop clustered voting to improve the message trust in vehicular ad-hoc networks. Int. J. Adv. Telecommun. Electrotechn. Signals Syst. **7**(1), 1 (2018). https://doi.org/10.11601/ijates.v7i1.245
5. Abbas, A.H., Audah, L., Alduais, N.A.M.: An efficient load balance algorithm for vehicular ad-hoc network. In: IEEE 2018, pp. 207–212 (2018)
6. Aissa, M., Belghith, A., Bouhdid, B.: Cluster connectivity assurance metrics in vehicular ad hoc networks. Proc. Comput. Sci. **52**, 294–301 (2015). https://doi.org/10.1016/j.procs.2015.05.088
7. Azizian, M., Cherkaoui, S., Hafid, A.S.: A distributed d-hop cluster formation for VANET. In: IEEE Wireless Communications and Networking Conference WCNC 2016, pp 1–6 (2016)
8. Bhosale, P., Vidhate, A.: An agglomerative approach to elect the cluster head in VANET. In: SCOPES 2016, pp 1340–1344 (2016)
9. He, Q., Yang, T.: A vehicular ad hoc networks clustering algorithm based on position-competition. In: CENet 2017, pp 1–8 (2017)
10. Cherkaoui, B., Beni-hssane, A., Erritali, M.: A clustering algorithm for detecting and handling black hole attack in vehicular ad hoc networks. In: Rocha, Á., Serrhini, M., Felgueiras, C. (eds.) Europe and MENA Cooperation Advances in Information and Communication Technologies, pp. 481–490. Springer International Publishing, Cham (2017). https://doi.org/10.1007/978-3-319-46568-5_49
11. Eze, E.C., Zhang, S.-J., Liu, E.-J., Eze, J.C.: Advances in vehicular ad-hoc networks (VANETs): challenges and road-map for future development. Int. J. Autom. Comput.Autom. Comput. **13**, 1–18 (2015)
12. Hadded, M., Muhlethaler, P., Laouiti, A., Saidane, L.A.: A novel angle-based clustering algorithm for vehicular ad hoc networks. In: Laouiti, A., Qayyum, A., Saad, M.N.M. (eds.) Vehicular Ad-Hoc Networks for Smart Cities, pp. 27–38. Springer, Singapore (2017). https://doi.org/10.1007/978-981-10-3503-6_3
13. Hashim, A.A., Shariff, A.R.M., Fadilah, S.I.: The modified safe clustering algorithm for vehicular ad hoc networks. In: IEEE 15th Student Conference on Research and Development (SCOReD), pp. 263–268 (2017)
14. Lee, J., Ahn, B.: An efficient message protocol using multichannel with clustering. Int. J. Appl. Eng. Res. **12**, 527–531 (2017)
15. Jalalvandi, S., Rafeh, R.: A cluster-based routing algorithm for V ANET. In: IEEE International Conference on Computer and Communications, pp. 2068–2072 (2016)
16. Kannekanti, S.Y., Nunna, G.S.P., Bobba, V.K.R., Yadama, A.K.: an efficient clustering scheme in vehicular ad-hoc networks. In: IEEE 2017, pp 282–287 (2017)

17. Kaur, A., Sindhwani, M., Arora, S.K.: Unity in togetherness: a review on clustering algorithms in vehicular ad- hoc networks. In: International Conference on Intelligent Circuits and System, pp 106–114 (2018)
18. Marzak, B., Toumi, H., Talea, M., Benlahmar, E.: Cluster head selection algorithm in vehicular ad hoc networks. In: IEEE 2015, pp 1–4 (2015)
19. Marzak, B., Toumi, H., Benlahmar, E., Talea, M.: Performance analysis of routing protocols in vehicular ad hoc network. In: El-Azouzi, R., Menasché, D.S., Sabir, E., De Pellegrini, F., Benjillali, M. (eds.) Advances in Ubiquitous Networking 2. LNEE, vol. 397, pp. 31–42. Springer, Singapore (2017). https://doi.org/10.1007/978-981-10-1627-1_3
20. Ouahou, S., Bah, S., Bakkoury, Z., Hafid, A.: Multi-hop clustering solution based on beacon delay for vehicular ad-hoc networks. In: Sabir, E., Armada, A.G., Ghogho, M., Debbah, M. (eds.) Ubiquitous Networking, pp. 357–367. Springer International Publishing, Cham (2017). https://doi.org/10.1007/978-3-319-68179-5_31
21. Prakash, R., Manivannan, P.V.: Simplified node decomposition and platoon head selection: a novel algorithm for node decomposition in vehicular ad hoc networks. Artif. Life Robot. 22(1), 44–50 (2016). https://doi.org/10.1007/s10015-016-0338-x
22. Qiu, T., Chen, N., Li, K., Qiao, D., Fu, Z.: Heterogeneous ad hoc networks: architectures, advances and challenges. Ad hoc Netw. 52, 1–10 (2016)
23. Rani, S., Ahmed, S.H.: Multi-hop network structure routing protocols. In: Rani, S., Ahmed, S.H. (eds.) Multi-hop Routing in Wireless Sensor Networks. SECE, pp. 45–58. Springer, Singapore (2016). https://doi.org/10.1007/978-981-287-730-7_4
24. Rasheed, A., Gillani, S., Ajmal, S., Qayyum, A.: Vehicular Ad Hoc network (VANET): a survey, challenges, and applications. In: Laouiti, A., Qayyum, A., Saad, M.N.M. (eds.) Vehicular Ad-Hoc Networks for Smart Cities, pp. 39–51. Springer, Singapore (2017). https://doi.org/10.1007/978-981-10-3503-6_4
25. Ren, M., Khoukhi, L., Labiod, H., Zhang, J., Veque, V.: A new mobility-based clustering algorithm for vehicular ad hoc networks (VANETs). In: IEEE/IFIP NOMS 2016, pp. 1203–1208 (2016)
26. Ren, M., Zhang, J., Khoukhi, L., Labiod, H., Vèque, V.: A unified framework of clustering approach in vehicular ad hoc networks. IEEE Trans. Intell. Transp. Syst.Intell. Transp. Syst. 19, 1–14 (2017)
27. Kandali, K., Bennis, L., Bennis, H.: A new hybrid routing protocol using a modified k-means clustering algorithm and continuous Hopfield network for VANET. IEEE Access 9, 47169–47183 (2021). https://doi.org/10.1109/ACCESS.2021.3068074
28. Alharthi, A., Ni, Q., Jiang, R.: A privacy-preservation framework based on biometrics blockchain (BBC) to prevent attacks in VANET. IEEE Access 9, 87299–87309 (2021). https://doi.org/10.1109/ACCESS.2021.3086225
29. Hossain, M.A., et al.: Multi-objective harris hawks optimization algorithm based 2-hop routing algorithm for CR-VANET. IEEE Access 9, 58230–58242 (2021). https://doi.org/10.1109/ACCESS.2021.3072922
30. Kazi, A.K., Khan, S.M., Haider, N.G.: Reliable group of vehicles (RGoV) in VANET. IEEE Access 9, 111407–111416 (2021) https://doi.org/10.1109/ACCESS.2021.3102216
31. Din, I.U., Ahmad, B., Almogren, A., Almajed, H., Mohiuddin, I., Rodrigues, J.J.P.C.: Left-right-front caching strategy for vehicular networks in ICN-based internet of things. IEEE Access 9, 595–605 (2021). https://doi.org/10.1109/ACCESS.2020.3046887
32. Funderberg, L.E., Ren, H., Lee, I.-Y.: Pairing-free signatures with insider-attack resistance for vehicular ad-hoc networks (VANETs). IEEE Access 9, 159587–159597 (2021). https://doi.org/10.1109/ACCESS.2021.3131189
33. Temurnikar, A., Verma, P., Choudhary, J.T.: A survey: routing protocol security algorithm and simulation tools in VANET. In: Conference INCETITDS-July -20 Ghaziabad UP (2020)

34. Temurnikar, A., Verma, P., Dhiman, G.: A PSO enable multi-hop clustering algorithm for VANET. Int. J. Swarm Intell. Res. **13**(2), 1–14 (2021). https://doi.org/10.4018/IJSIR.2022040 1.oa7
35. Temurnikar, A., Verma, P., Choudhary, J.: Development of multi-hop clustering approach for vehicular ad-hoc network. Int. J. Emerg. Technol. **11**(4), 173–177 (2020)
36. Temurnikar, A., Verma, P., Choudhary, J.T., et al.: Securing vehicular ad hoc network against malicious vehicles using advanced clustering technique. In: 2nd IEEE International Conference on Data Engineering and Application, IDEA-2K20 (2020)
37. Temurnikar, A., Verma, P., Choudhary, J.T.: Vehicular ad-hoc network security and data transmission: survey and discussions. ,JETIR, June-2019, ISSN 2349–5162 (2019)
38. Temurnikar, A.: Security Issues of Vehicular Ad hoc Network. Published in International Conference on Emerging Trends in Technology and Science, People's University Bhopal (2018)
39. Temurnikar, A., Sharma, S.: Secure and stable VANET architecture model. IJCSN J. **2**(1), 37–43 (2013)

An Improved Deep Learning Model Implementation for Pest Species Detection

Nikita Agarwal, Tina Kalita, Ashwani Kumar Dubey[(⊠)], Shreyas Om,
and Anika Dogra

Amity School of Engineering and Technology, Amity University
Uttar Pradesh, Noida, Uttar Pradesh, India
dubeylak@gmail.com

Abstract. Pests account for more than half of all known animal species and can be found in all types of environments. These pests are one of the main reasons behind the decline in crop yield. Accurate recognition of pests must be done so that timely measures can be taken according to the type of pest and the losses are reduced over time. As a result, Deep Learning has been utilized to identify pest species more quickly and accurately. In this paper, a Vision Transformer (ViT) based model is being used to detect the pest species more accurately. A large data set is being used containing 50 different species of insect pests. It incorporates well over 34,089 photos that are divided into 50 categories and have a naturally long-tailed distribution. A comparison of the performance of ViT is done with the ResNet model on the same data and it is found that the ViT model performs better than the ResNet model.

Keywords: Deep learning · Pest Species Recognition · ResNet Vision transformer

1   Introduction

Pest can be considered as a type of natural disaster. This affects how a plant grows and contributes to the death of the plant [1]. One of the key elements influencing the productivity of agricultural products is pests [2]. Forest pests and diseases influence forest fuels and wildfire, as they kill plants, hence increasing their vulnerability to ignition and catastrophic wildfire. If productivity gets challenged, it will affect us humans directly or indirectly and as for a country like India which is mostly dependent on agriculture, the impact will be huge [3]. To determine the yield and the quality of plants, the recognition of plant diseases and pests is very important [4, 5]. Therefore, prompt preventive measures to avert financial losses are made possible by the precise identification of insect pests. This is the major reason behind the aim of our paper, to identify pests of all kinds so that harmful pests can be eradicated.

The manual detection of pests is being done by human labor, which is not very accurate [6, 7]. Automation is necessary in this area since it will accurately identify pests in crops [8]. Keeping this in mind, our paper will be based on the identification of pests using deep learning. In the paper, we have used a pre-trained model to process the dataset that we collected and perform various processes like data augmentations and more on the dataset.

It can be challenging to draw out the complex and complicated picture information of the feature using traditional manual image classification and identification methods, which can only extract the fundamental features. When comparing the image recognition methods, it is found that those based on deep learning do not draw out the specific features but rather find appropriate features through iterative learning. Using this we get contextual and global features. The accuracy for recognition increases with robustness. Thus, obstruction can be removed by using the deep learning approach. Unsupervised learning can be done straight from the actual image to get information about many image feature levels, including low-level, middle, and elevated semantic features.

Traditional algorithms for detecting plant diseases and pests primarily use manually constructed features for picture recognition, which is challenging and relies on luck and experience. The task of extracting and automatic learning using these algorithms is difficult. So, when it comes to deep learning automatic extraction of features is done from huge amounts of data. The model, which has several layers and high feature expression and autonomous learning capabilities, can autonomously extract features of an image for image identification and classification. Hence, we can conclude that the potential of deep learning is enormous for the recognition of pests. Hence, this is the reason why our proposed work uses Deep Learning instead of Machine Learning.

Our contributions to this paper have been summarized as follows:

- Based on our knowledge, we have built a deep-learning model that detects the pest species with the help of the IP102 dataset.
- We sought to achieve high accuracy and prediction for the same.
- The model can predict various species of pests such as rice leaf roller, paddy stem maggot, brown plant hopper, etc. The list of some of the pest classes can be seen in Table 1. The model covers 50 species of pests.

2 Related Work

Much work has been done with different datasets and using different architectures by other researchers having the same aim as what has been stated by us. There have been researches with a dataset which has a wide range of pests found in the forest, it is also found to meet the requirements for both experimental and natural environments, the dataset also included pests that were from different periods of time and had not been included previously. Various algorithms for object detection were used on their dataset. With the faster RCNN two-stage approach is used, by sliding window the feature maps are scanned, and classified and the information about the coordinates is regressed. For YOLOV4 and SSD, a one-stage approach is used. The information about the location and the category is regressed. Furthermore, the assessment of Deformable DETR was done, which is an end-to-end transformer-based object detection algorithm [1].

In [16] it was proposed that Support vector machine, artificial neural network, and convolutional neural network are three classifiers that are assessed for classification. For experimenting, OpenCV and Keras libraries in Python are used. The Polynomial kernel in SVM (87.6%) and the 5-6-2 architecture in ANN (94.7% validation accuracy and 93.2% test accuracy) both produce the highest levels of accuracy. With an input layer of dimensions $250 \times 250 \times 3$, CNN is used and performs admirably. Due to comparable patterns and lighting circumstances, the study notes difficulties in telling crops and weeds apart, and it makes suggestions for future improvements in feature extraction to increase classifier performance.

[17] In order to obtain reflectance values from the red, green, and blue bands of RGB photographs for the research on an Australian chili farm, image processing techniques were used. Then, using machine learning methods like Random Forest (RF), Support Vector Machine (SVM), and K-Nearest Neighbours (KNN), these attributes were employed to detect weeds. The highest accuracy for weed detection was attained by RF (96%), followed by SVM (94%), and KNN (63%). In the study, RF and SVM showed promising results in weed recognition using UAV photos, demonstrating the usefulness of image processing and machine learning for early weed detection in agricultural fields.

It was concluded in [9], Color (CH) features performed badly on most evaluation metrics for handcrafted features when compared to texture features. This suggested that when insect pests appear in the wild, texture features become more important. In terms of deep features, the KNN outperforms the SVM classifier overall. The KNN results perform better than the SVM on the majority of metrics, especially when using AlexNet features.

To determine whether or not an image's classification was accurate, two generalized linear models were employed with binomial distribution in this paper. Only species identification was utilized as an explanatory variable in Model 1. In model 2, from the output layer they used the top 1 value in CNN and the as a measure of body size, image size in megapixels was extracted with the help of exiftool v.11.06 via the exif r-package from image metadata (exif) (hereafter referred to as body size). To isolate the impacts of the three explanatory variables on the prediction, a sensitivity analysis was done for model 2 [10].

In this research, an effective pest recognition model called ExquisiteNet was suggested. Essentially, ExquisiteNet is made up of two components. Max feature expansion block (ME block) and dual fusion with squeeze-and-excitation-bottleneck block (DFSEB block) are the two. A benchmark dataset for pests called IP102 was used to test the model's performance. Without any data augmentation, the model achieves a higher precision of 52.32% on the IP102 test set [11].

This research suggests an automated pest recognition technique centered on Vision Transformer (ViT) to increase the precision of computer categorization of plant diseases and pests. The insect pest and plant conditions data sets are improved using techniques like Histogram Equalisation, Laplacian, Gamma Transformation, etc. in order to prevent training overfitting. After that, train the created ViT neural network using the improved data set to achieve automatic categorization of plant illnesses and insect pests. According to the simulation findings, the built-in ViT model achieves a test recognition success level of 96.71% on the publicly available Plant_Village data set for plant diseases and insect pests, which is around 1.00% higher compared to the plant pathogen and pest recognition approach based on conventional CNN [12].

3 Methodology

The methodology of our proposed work is as follows:

3.1 Dataset

We have used IP102, a large-scale dataset for insect pest recognition. The IP102's taxonomy is hierarchical and grouping of insect pests into the same high-level category is done according to whichever agricultural product they affect [9] (Fig. 1).

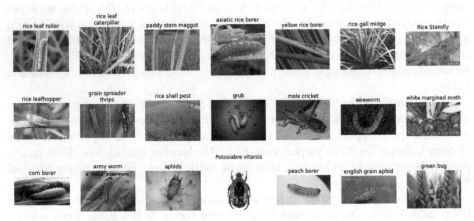

Fig. 1. A snippet of some of the pest species from IP102 dataset [9]

The study makes use of a sizable dataset known as IP102, which includes data from 50 distinct species groups. The total number of images in this varied dataset is over 75,000, with a maximum of 11,227 images per category. Following a roughly 6:1:3 split ratio at the sub-class level, the dataset is methodically separated into three subsets for training, validation, and testing. Specifically, the validation set includes 7,508 photos, the testing set includes 22,619 images, and the training set includes 45,095 images.

The IP102 dataset includes numerous photos as well as useful information about each occurrence. These characteristics offer important context and details about the photographs. The attributes include "Crop," which refers to the agricultural product most commonly impacted by the insect pest, "Location," which specifies the location where the image was taken, "Season," which indicates the particular season during which the image was taken, and "Life Stage," which describes the stage of the insect pest's life cycle that is depicted in the image. These characteristics add value to the information and may help researchers better understand how different insect species interact and affect agriculture.

3.2 Deep Learning

Deep learning models for computer vision that are frequently employed are convolutional neural networks (CNNs). Convolutional layers are used to extract regional characteristics, such as edges and textures, and pooling layers are used to reduce the size of the spatial domain while preserving crucial data. CNNs effectively identify patterns and objects in images, making them useful for tasks like object and image classification.

Deep learning models called Vision Transformers (ViTs) were created for image identification. Contrary to CNNs, ViTs use transformer topologies to interpret images as sequences of fixed-size patches. To identify global context and dependencies in visuals, they employ self-attention. Transfer learning is made easier by ViTs' parameter efficiency, interpretability, and scalability for different image sizes without architectural changes. For the architecture of Vision Transformers refer to Fig. 3.

The processing and design of CNNs and Vision Transformers are different. In contrast to ViTs, which use patches and self-attention to model relationships, CNNs convolve filters across the image grid. Large visuals and varied input dimensions work particularly well with ViTs, which also offers interpretability through attention maps. Smaller datasets and spatially localized characteristics continue to be well-suited for CNNs. Applications of computer vision benefit greatly from both ViTs and CNNs.

Another kind of model covered in these segments are ResNets. Deep learning models called Residual Networks (ResNets) are frequently employed in computer vision. By incorporating skip connections or shortcuts, they solve the vanishing gradient issue and allow the network to learn residual mappings. These shortcuts enable easy and accurate training of very deep networks by enabling direct information transfer between layers. Refer to Fig. 4 for the architecture of ResNet.

Convolutional neural networks (CNNs) and vision transformers (ViTs) are not comparable to ResNets in terms of their architectural design and training methodology. While ResNets use skip connections to get over the degradation issue in deep architectures, CNNs use convolutional layers for grid-like image processing. In contrast, ViTs use self-attention mechanisms to perceive images as collections of fixed-size patches. Skip connections in ResNets make it easier to train deep models, whereas ViTs concentrate on identifying long-range dependencies in images. Each model has its advantages, and depending on the complexity of the task at hand and the size of the network, different applications can be made of them (Fig. 2).

3.3 Data Augmentation

This method is crucial for raising the generalisation capabilities of the model while lowering the risk of overfitting. By applying a variety of augmentation techniques, the researchers are able to successfully increase the dataset's size and diversity, allowing the deep learning model to recognise trustworthy characteristics and patterns from a wider range of altered photos. Numerous geometric modifications, including rotation, vertical reflection, left-to-right translation, top-down translation, and horizontal reflection, are used as part of the data augmentation process in this study. These modifications greatly increase the diversity of the training data, enabling the model to recognise pests in a wide range of postures, viewpoints, and orientations, just like it might in the real world [13].

By exposing the model to a wider range of enhanced images during training, it is possible to achieve more generality and capture of the innate traits that identify different pest species. As a result, the model is better equipped to handle scenarios where the characteristics of pests might vary substantially in look and setting. A more balanced

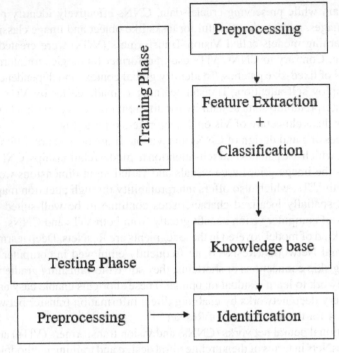

Fig. 2. Block diagram of Deep Learning

training dataset is produced through data augmentation, which also adds additional cases for the minority classes, addressing any potential data imbalances across different pest species. In general, data augmentation works as an effective technique to increase the model's adaptability and performance, allowing it to recognise and detect pest species with greater realism and accuracy [13].

3.4 Model Architecture

A comparison was made between Vision Transformer and ResNet50 to see which one of the two architectures is best suited for our aim.

A methodology for classifying images is called Vision Transformer, or ViT. Here, to the selected areas of the image, a Transformer-like design is applied. This pre-trained model is being used in our paper. A picture that has been divided into patches of fixed size, embedding of each one linearly, addition of embeddings that are positional and at last the vectors assembled being fed to conventional Transformer encoder that has been created by a sequence of vectors.

Recently, the Vision Transformer (ViT) has been recognized as an effective Option [14] to CNN i.e. Convolutional Neural network. CNNs are used widely for state-of-the-art computer vision and in tasks related to recognition of image [15]. When it comes to accuracy and efficiency, the performance of ViT is nearly 4 times better than that of CNN.

The vision transformer model's overall architecture entails-

- splitting an image into patches i.e., fixed sizes.
- Image patch flattening and creation of lower-dimensional linear embeddings from these flattened image patches.
- Embedding that is positional is then included.
- The state-of-the-art transformer encoder is then fed with the sequence as its input.
- Pre-training of the ViT model along with image labels is done before fully supervised training on a large database.

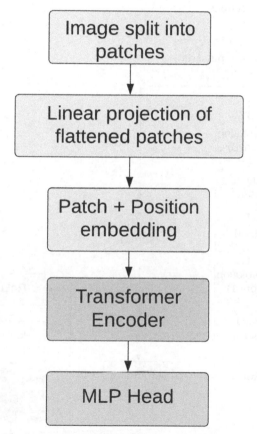

Fig. 3. Architecture of the Vision Transformer (ViT)

Finally, the fine-tuning of image classification with the downstream dataset is done.

Another architecture used by us is ResNet. Shaoqing, Xiangyu Zhang, Ren, Kaiming He and Jian Sun introduced the Residual Network (ResNet), one of the popularly known deep learning models, in their study.

One convolution and pooling phase is succeeded by four layers of identical behaviour to make up the ResNet. By each layer, the pattern is followed exactly as before. After two convolutions, the input is skipped and the convolution of 3×3 with the dimensions of the feature map fixed [64, 128, 256, 512] correspondingly. To add to this, the height and weight of an entire layer remain the same.

The issue related to training of very deep networks is being resolved, because of the advent of Residual blocks. These blocks build the ResNet model. The infamous vanishing gradient is one of the issues that ResNets address. This is because when the platform is excessively deep, the gradients used to compute the loss function simply drop to zero after a number of chain rule applications.

As a result, there is no learning taking place because the weights' values are never updated. With ResNets, gradients from the previous layer to initial filters can flow straight through the skip connections.

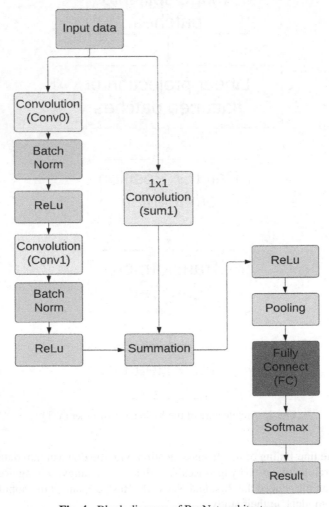

Fig. 4. Block diagram of ResNet architecture

3.5 Classification

The aforementioned architectures received the IP102 dataset as input, and each architecture carried out its corresponding tasks. We began the testing phase after completing the dataset's training and validation. Finally, we anticipated that our algorithm would accurately categorize the dataset of 50 pest species. The details of this approach's results are in the results section.

4 Results

In this study, we employed two powerful models, ResNet50 and Vision Transformer, to conduct a comparative analysis of their performance on our dataset which included almost 50 different species of insects.

Out of the these utilized in our paper, a few of them are listed in Table 1 below.

Table 1. List of some of the classes from the IP102 dataset

Label	Name
1	Rice Leaf Roller
2	Rice Leaf Caterpillar
3	Paddy Stem Maggot
4	Asiatic Rice Borer
5	Yellow Rice Borer
6	Rice Gall Midge
7	Rice Stemfly
8	Brown Plant Hopper
9	White Backed Plant Hopper
10	Small Brown Plant Hopper
11	Rice Water Weevil
12	Rice Leafhopper
13	Grain Spreader Thrips
14	Rice Shell Pest
15	Grub
16	Mole Cricket
17	Wireworm
18	White Margined Moth
19	Black Cutworm

Figures 5 and 6 show the results of the testing dataset being loaded on Vision transformer (ViT) and ResNet50 respectively. For Fig. 5, utilizing Vision Transformer (ViT) model, it demonstrated us with reliable and accurate predictions for several species, including Flax Budworm, Rice Leaf Roller, Rice Gall Midge, Limacodidae, Scirtothrips Dorsalis Hood, Blister Beetle, and Wheat Blossom Midge (a to h, respectively). Similarly, in Fig. 6, ResNet50 displayed accurate predictions for Legume Blister Beetle, Aphids, Army Worm, Red Spider, Prodenia Litura, Panonchus Citri Mcgregor, and Aphids (a to f, and h, respectively). However, it should be highlighted that for species (g) in Fig. 6, ResNet50 misclassified it as "Aphids" rather than its true name, which is "Tarnished Plant Bug."

(a)   (b)   (c)   (d)

(e)   (f)   (g)   (h)

Fig. 5. Results obtained when using Vision Transformer

(a)   (b)   (c)   (d)

(e)   (f)   (g)   (h)

Fig. 6. Results obtained when using ResNet50

We looked at the models' training and testing accuracies throughout eight epochs to further evaluate and compare them, as shown in Table 2. ResNet50 shown a considerable increase in training accuracy, going from 59.8% to 89.23%. Its testing precision, however, only slightly improved from 66.8% to 73.11%. However, Vision Transformer outperformed the competition throughout training, increasing its initial accuracy to an astonishing 95.4% from a higher beginning accuracy of 83.8%. Additionally, its testing accuracy significantly improved, going from 73.7% to a stunning 92.8%.

Based on these results, the superiority of Vision Transformer becomes evident, as it achieved a final testing accuracy of 92.8% compared to ResNet50's 73.11%. This highlights the efficacy of the Vision Transformer model for accurately classifying the species within our dataset, making it a more satisfactory choice for this specific task.

Table 2. Training and Testing Accuracy of Resnet and Vision Transformer for 8 Epochs

ResNet (Training)	ResNet (Testing)	ViT (Training)	ViT (Testing)
0.598	0.668	0.838	0.737
0.6722	0.8435	0.722	0.598
0.8696	0.8266	0.898	0.804
0.8870	0.9677	0.912	0.852
0.8435	0.8132	0.8312	0.883
0.9281	0.8718	0.802	0.905
0.8712	0.7922	0.878	0.919
0.8923	0.7311	0.954	0.928

Obtaining a real-time dataset of insect pest species proved to be a considerable challenge over the course of our endeavour. Because they are so sensitive to changes in the environment and the weather, insects provide significant difficulty when it comes to precisely tracking their occurrences and whereabouts. It became difficult to acquire current and accurate data because of their erratic movements and fluctuating presence. Since these insect populations are dynamic, it was challenging to keep a complete and up-to-date dataset, which could have had an impact on the accuracy and generalizability of our models.

A significant danger that may affect the results of our endeavour was also discovered. It was centred on how photos from various insect pest classifications were distributed across the dataset. The training process could be unbalanced if the dataset is not properly distributed, and one class has a disproportionately high number of photos compared to the others. A biased model that performs better for that class may be produced by having the class with the most photos dominate the training and testing phases. As a result, the model's performance on underrepresented classes may be harmed, which would reduce its overall accuracy and resilience.

For our classification assignment to produce fair and trustworthy results, addressing this problem through thorough data pretreatment and augmentation procedures became essential.

5 Conclusion

The IP102 dataset was utilised in this study to examine the effectiveness of two model architectures, Vision Transformer (ViT) and Convolutional Neural Network (CNN), in identifying pest species. The outcomes unmistakably demonstrated that ViT outperforms ResNet50, a kind of CNN. ResNet50 only managed a meagre 73.11% test accuracy compared to ViT's astounding 92.8%. This notable performance difference demonstrates how much better Vision Transformer is in detecting pest species than traditional CNNs.

The outcomes demonstrate the potential of Vision Transformers as a reliable CNN substitute for computer vision applications, particularly for datasets with numerous intricate models, such as the IP102 dataset. ViTs outscored ResNet50 by a factor of more than four, attaining great accuracy and efficiency. The effectiveness of Vision Transformers in difficult real-world agricultural circumstances where precise pest species identification is necessary to reduce crop production losses is explained in this article.

This study underscores the necessity for contemporary deep-learning techniques while advancing the field of agricultural pest identification by offering insights into model performance and architectural choices. ViTs are promising solutions, but more research is needed to fully understand their potential in diverse computer vision applications. This will open the door for more precise and effective models that can aid in a variety of other areas besides only identifying pest species.

References

1. Liu, B., Liu, L., Zhuo, R., Chen, W., Duan, R., Wang, G.: A dataset for forestry pest identification. Front. Plant Sci. **13**(857104), 1 (2022). https://doi.org/10.3389/fpls.2022.857104
2. Li, W., Zheng, T., Yang, Z., Li, M., Sun, C., Yang, X.: Classification and detection of insects from field images using deep learning for smart pest management: a systematic review. Eco. Inform. **66**, 101460 (2021). https://doi.org/10.1016/j.ecoinf.2021.101460
3. Agarwal, A., Sarkar, A., Dubey, A.K.: Computer vision-based fruit disease detection and classification. In: Tiwari, S., Trivedi, M.C., Mishra, K.K., Misra, A.K., Kumar, K.K. (eds.) Smart Innovations in Communication and Computational Sciences. AISC, vol. 851, pp. 105–115. Springer, Singapore (2019). https://doi.org/10.1007/978-981-13-2414-7_11
4. Liu, J., Wang, X.: Plant diseases and pests detection based on deep learning: a review. Plant Methods **17**(1), 1–18 (2021). https://doi.org/10.1186/s13007-021-00722-9
5. Zhang, S., Jing, R., Shi, X.: Crop pest recognition based on a modified capsule network. Adv. Intell. Comput. Theor. Appl. (ICIC) **10**(1), 552–561 (2022). https://doi.org/10.1080/21642583.2022.2074168
6. Patel, P.P., Vaghela, D.B.: Crop diseases and pests detection using convolutional neural network. In: IEEE International Conference on Electrical, Computer and Communication Technologies (ICECCT), pp. 1–4 (2019). https://doi.org/10.1109/ICECCT.2019.8869510
7. Waleed, A., Momina, M., Saleh, A.: Custom CornerNet: a drone-based improved deep learning technique for large-scale multiclass pest localization and classification. Complex Intell. Syst., 1–18 (2022). https://doi.org/10.1007/s40747-022-00847-x
8. Kumar, Y., Dubey, A.K., Jothi, A.: Pest detection using adaptive thresholding. In: International Conference on Computing, Communication and Automation (ICCCA), pp. 42–46 (2017). https://doi.org/10.1109/CCAA.2017.8229828

9. Wu, X., Zhan, C., Lai, Y.-K., Cheng, M.-M., Yang, J.: IP102: a large-scale benchmark dataset for insect pest recognition. In: Conference on Computer Vision and Pattern Recognition (CVPR), pp. 8779–8788. IEEE (2019). https://doi.org/10.1109/CVPR.2019.00899

10. Hansen, O.L.P., et al.: Species-level image classification with convolutional neural network enables insect identification from habitus images. Ecol. Evol. **10**(2), 737–747 (2020). https://doi.org/10.1002/ece3.5921

11. Liu, L., Wang, R., Xie, C., Yang, P., Wang, F., Sudirman, S.: PestNet: an end-to-end deep learning approach for large-scale multi-class pest detection and classification. IEEE Access **7**, 45301–45312 (2019). https://doi.org/10.1109/ACCESS.2019.2909522

12. Chen, C.-J., Huang, Y.-Y., Li, Y.-S., Chen, Y.-C., Chang, C.-Y., Huang, Y.-M.: Identification of fruit tree pests with deep learning on embedded drone to achieve accurate pesticide spraying. IEEE Access **9**, 21986–21997 (2021). https://doi.org/10.1109/ACCESS.2021.3056082

13. Kasinathan, T., Singaraju, D., Uyyala, S.R.: Insect classification and detection in field crops using modern machine learning techniques. Inf. Process. Agric. **8**(3), 446–457 (2021). https://doi.org/10.1016/j.inpa.2020.09.006

14. Chen, W., et al.: A simple single-scale vision transformer for object localization and instance segmentation. In: Avidan, S., Brostow, G., Cissé, M., Farinella, G.M., Hassner, T. (eds.) Computer Vision – ECCV, pp. 711–727. Springer, Cham (2021). https://doi.org/10.1007/978-3-031-20080-9_41

15. Dosovitskiy, A., et al.: An image is worth 16x16 words: transformers for image recognition at scale. arXiv preprint arXiv:2010.11929, pp. 1–21 (2020). https://doi.org/10.48550/arXiv.2010.11929

16. Sarvini, T., Sneha, T., Sukanya Gowthami, G.S., Sushmitha, S., Kumaraswamy, R.: Performance comparison of weed detection algorithms. In: 2019 International Conference on Communication and Signal Processing (ICCSP), Chennai, India, pp. 0843–0847 (2019). https://doi.org/10.1109/ICCSP.2019.8698094

17. Islam, N., et al.: Early weed detection using image processing and machine learning techniques in an Australian chilli farm. Agriculture **11**, 387 (2021). https://doi.org/10.3390/agriculture11050387

Identification of Diseases Affecting Mango Leaves Using Deep Learning Models

Thaseentaj Shaik and Sudhakar Ilango Swamykan[(⊠)]

VIT-AP University, Amaravati, Vijayawada 522237, India
{shaikthaseentaj.20phd7170,Sudhakar.ilango}@vitap.ac.in

Abstract. Agriculture is a country's economic backbone. Agriculture is quickly expanding in India. Current technology raised food yields and quality, but leaf diseases and raised pesticide use have lowered agricultural production. Mango is one of the world's most popular fruits, particularly in India. Many pathological problems, severe illnesses, and pesticide use significantly reduce mango crop yields. Mango leaf diseases have a significant influence on mango quality and output. It is difficult to diagnose mango leaf disease with the naked eye so it is very important to develop a model that can detect leaf diseases at early stages. Previously, researchers used a number of computer-aided and machine-learning approaches to classify mango leaf diseases. According to reports, these procedures have various downsides. In existing research, they only focused on 3 to 4 diseases some researchers only focused on fungal diseases and some researchers only focused on bacterial diseases. In this research, The aim is to detect 13 different fungal and bacterial mango leaf diseases using CNN techniques. The photographs were captured in the Andhra Pradesh state's Chittoor district, which is home to India's largest mango-growing region. The dataset consists of 1100 images. The dataset was trained using popular CNN approaches, namely GooGLeNet, EfficientNet, and ResNet-50. The results indicate that the EfficientNet model achieved the highest accuracy of 98.7% in classifying mango leaf diseases. The study demonstrates the potential of deep learning models in identifying diseases affecting mango leaves, which could aid in timely disease detection and control.

Keywords: Mango leaf diseases · deep learning · Convolutional Neural Network · EfficientNet · GooGLeNet · ResNet-50 · disease detection · agriculture production

1 Introduction

Agriculture in India plays an important role in feeding both human and animal populations globally. Mango is special in that it is easily cultivable due to its ability to adapt to many elements such as climate, soil, and location. Nearly 18.1% of India's Gross domestic product in the agriculture division is accounted for by mango. Mangoes are regarded as a vital agricultural product in India,

S. Tiwari et al. (Eds.): AI4S 2023, CCIS 1907, pp. 132–144, 2023.
https://doi.org/10.1007/978-3-031-47997-7_10

and India accounts for around 52.63% of total global output. In India, Andhra Pradesh is the second largest mango producing state. Figure 1 describes Year-wise Andhra Pradesh's mango production from 2001 to 2021.

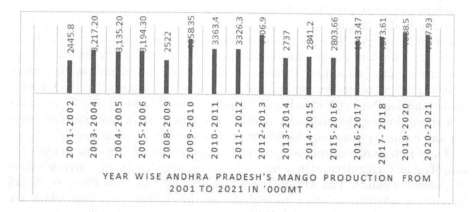

Fig. 1. Year wise Andhra Pradesh's mango production from 2001 to 2021 in '000MT (Data collected from NHB National Horticultural Board [11]).

Despite this, mango producers face a slew of problems, resulting in a negative growth rate. Many variables influence mango production, ranging from bacterial and fungal infections to climate change and soil infertility. Yet, the main cause of this slump is the farmers' inability to identify the illnesses that harm mango plants. Microbes such as fungi, parasites, bacteria, and algae result in illnesses in mangoes at all stages of development, from seedling through fruit consumption. Some illnesses are caused by dietary deficits as well as unknown aetiologies [1]. Mango fruit disorders are illnesses brought by outside forces and manifest as physical defects like blemishes on the fruit's skin. This lowers the price and lowers the quality. The illnesses have the capacity to present as various forms of deformities, necrosis, mildew, rot, stem bleeding, and wild, sooty mold. India's growers are particularly concerned about bacterial canker, anthracnose, sooty mold, and phoma blight, blacktip, dieback, mildew, deformity. Preventive measures are the only way to effectively reduce losses caused by the majority of mango diseases. Consequently, it is essential to spot infections early on in order to protect the trees, avoid any financial losses, and boost agricultural output for the farmers.

2 Literature Survey

In their research, Mustafa Merchant and colleagues [18] developed a clustering technique to detect nutrition-related deficiencies in mango trees using high-definition camera images of mango leaves. This involved extracting and comparing RGB and texture characteristics of the images, along with preprocessing techniques such as rescaling contrast enhancement for better feature extraction. To categorize the leaves as healthy or diseased The k-means clustering algorithm was used, and further classify leaves as deficient in Potassium, Iron, Nitrogen, or Copper. This approach can help farmers detect deficiencies early and take countermeasures to prevent unhealthy plant growth.

Uday Pratap Singh and his team [26] presented a model named "Multilayer Convolution Neural Network" to categorize mango leaf photos as healthy or unhealthy with Anthracnose fungus infection. This methodology, which includes machine learning and computer vision techniques including classification, pattern recognition, and object extraction, was contrasted with others like PSO, Radial Basis Function Neural Network and Support Vector Machine. 97.13% accuracy was attained using the suggested model. These studies show how machine learning and computer vision are useful for predicting early plant diseases.

Only high-resolution photos plus an "Artificial Neural Network model" can identify minor irregularities in plant photographs, allowing for the early detection of illnesses in plants. The APGWO was suggested by Tan Nhat Pham et al. [20] for feature selection in order to find the pertinent characteristics, which are subsequently fed into the model. With an accuracy rate of 89.14%, our method surpassed other CNN designs including ResNet-50, VGG16, and Alex net.

Sunayana Arya et al. [2] conducted a thorough analysis and came to the conclusion that AlexNet performed better given the dataset utilized. The AlexNet architecture model outperforms other designs like GoogleNet, DenseNet, ResNet, VGGNet and SqueezeNet in terms of total system performance.

S Arivazhagan et al. [1] proposed an automated "deep learning-based" system for diagnosing mango leaf diseases such as Alternaria leaf spots, Leaf Gall and Leaf burn, Anthracnose, and Leaf Webber. The data set used in this approach [5] consists of a total of 1200 images from which 600 photos are used for the training process, and 600 images are used for the testing process. The system has demonstrated its viability for use in real-time applications by achieving a high accuracy of 96.67% in identifying leaf illnesses in mango trees.

Another solution suggested by B Prakash et al. [22] employs the K-means clustering technique for classification and the BPNN technique for segmentation. The difficulty of manually detecting and identifying mango leaf disease is eliminated by this approach. The suggested technique is evaluated using various cluster sizes and test datasets, creating the top-performing mango leaf disease diagnosis and control prediction system with an accuracy of 94%.

An algorithm [27] created by Meenakshi Sood et al. can categorize capsicum plants as healthy or diseased with fungal or bacterial illnesses like bacterial canker or anthracnose. It can identify diseases that affect capsicum plants. On various portions of the plant, such as the stem or leaves, these diseases' symptoms might be seen. Using a Decision Tree, K-Nearest Neighbour, and Support Vector classifiers, Linear Discriminant classification is carried out. K-means clustering is utilized to extract features, such as the diseased region on the leaf. The KNN and SVM classifiers, both of which had a 100% accuracy rate, produced the best results.

Junde Chen et al. [12] suggested a transfer learning method employing a pre-trained technique from the ImageNet dataset, which they named INC-VGGNet, to identify photos of sick plant leaves. In comparison to other cutting-edge methods like ResNet-50, DenseNet-201, and Inception V3, this strategy displayed a considerable performance gain, obtaining a validation accuracy of 92%. The PlantVillage dataset was used to test the proposed deep learning architecture, demonstrating both the viability of the method and its effectiveness in detecting plant illnesses.

According to Swetha K et al. [28], detecting infection in mango leaves involves identifying areas of anthracnose infection through a thresholding segmentation methodology that takes into account the size and intensity of the spots. To accurately assess the affected region, traits such as the area and perimeter of the afflicted sections are also considered in addition to thresholding. Spot intensity has been used as a metric in several existing approaches to predict the percentage of infected areas

In contrast to manual methods, Md. Rasel Mia et al. [19] have developed a "neural network ensemble model" for detecting mango leaf diseases that aid in accurate disease identification. This model uses machine learning to monitor symptoms of different types of leaves, including "Dag disease", "Golmachi disease", "Moricha disease", and "Shutimold", and generates training data through classification techniques using images of disease-infected leaves. The suggested model can categorize the investigated illness with an average combined accuracy of 80%, allowing for primary detection and treatment of the affected leaves with essential therapies, thus improving mango production.

Using a dataset of plant leaves, Konstantinos P. Ferentinos created a method for detecting and diagnosing plant diseases using a variety of deep learning approaches, including a convolutional neural network (CNN) model [14]. After training many CNN model architectures, including AlexNet and GoogLeNet, VGG CNN was able to successfully detect the appropriate healthy plant or disease combination from the straight-forward plant leaf photos with a success rate of 99.53%. These findings show that CNN models are ideal for automating the identification and diagnosis of plant diseases.

Parul Sharma et al. [25] investigated a possible cure for overfitting by linking the performance of two methods, one trained using segmented leaf images (S-CNN), and the another with full images (F-CNN). The accuracy rate of the S-CNN model doubled to 98.6% when evaluated on unseen data with ten illness classifications in comparison to the F-CNN model's performance. This study is a great illustration of an automated tool that non-experts may use to quickly identify plant leaf diseases.

An interdisciplinary subject called computer vision (CV) has been extensively applied in a number of fields, including the identification of tree illnesses [3–10, 16, 17, 24, 30]. Recent years have seen the effective integration of computer vision and image processing technology to deliver an excellent answer for plant disease early detection. Using images from the Plant Village dataset [13], Kien Trang et al. [29] developed a deep neural network method to identify mango leaf diseases.

A range of pre-processing methods, such as rescaling, contrast enhancement, and centre alignment, was used to preprocess the images. The accuracy rate of 88.46% attained by the suggested model-higher than that of other architectural models like MobileNetv2, AlexNet, and Inceptionv3-is the consequence of the incorporation of Transfer Learning, which provides a considerable benefit in the learning process.

2.1   Disease Affecting Mango Leaves

Agriculture production typically declines as a result of plant diseases. The majority of fungal diseases affect plants' leaves, as compared to other types of illnesses [21]. No matter what section, the stem, fruits, vegetables, fruits, and all items are affected. These diseases' primary contributing elements can be divided into two categories: disorder and disease. Diseases are brought on by things like bacteria, fungi, or algae, whereas disorders are brought on by variables like nutritional shortage, soil moisture content, rainfall, temperature, etc. Mango leaves frequently have various illnesses. [15] If the cultivation is carried out at a lower rate, these diseases are detected. Farmers that cultivate wider areas often struggle to detect infections, which has a negative impact on their crops and causes increased crop loss. Farmers use a lot of insecticides to counter this effect, endangering human lives in the process [23].

2.2   Factor Influencing Fungal Diseases

These disorders' primary contributing elements can be generally divided into two categories. Sickness and disorder are them. Bacteria, fungi, and algae are some of the different ways or sources that spread disease. In earlier parts, a concise overview was provided of the various fungal attacks on trees and the diseases that resulted from them. The remaining elements that contribute to the disease are discussed in more detail in this section. The top four items taken into account here that the climatic elements would affect are;

Temperature - The two key characteristics of the fruit that are influenced by this aspect are fruit quality and maturity. Mango fruit ovules abort when the temperature is between 12 °C to 44 °C. The stability of the entire tree, which can survive the deteriorating nature of the leaves, depends on the temperature. The mango trees, foliage, and fruits are not impacted by this temperature, however, this varies from region to region and cannot be regulated to a specific degree.

Rainfall - This element has an impact on the stability of the tree's whole structure, including its leaves and fruits. This aspect has a significant impact on how developed the leaves and fruits will be. Growing is considered successful when the range of rainfall is 75 cm to 350 cm. But, it shouldn't rain throughout the fruit-set and blooming seasons. The entire set of production is directly impacted when there is water logging since the leaves begin to spoil. During at least four months, the weather is predicted to stay dry, allowing for greater concentration on flowering and harvest.

Light -This factor plays a key role in the development of fruit and leaves. Photosynthesis, which determines how many carbs will be accumulated, is solely accountable for this. Increased leaf production directly correlates with the production of fruits when there is sufficient light. It is stated that there should be sufficient light, not too much.

Nutrients - Young trees must have adequate nutrition to grow quickly and produce blossoms and fruits. Yet, when nutrients, like nitrogen, are consumed in excess, the leaf-to-fruit ratio is impacted, which alters the color of the fruit. All three sections are susceptible to illness when the ratio rises since the severity of diseases also rises.

3 Dataset Description

The dataset used in this research was collected from mango farms located in Chittoor district, Andhra Pradesh, India. Andhra Pradesh is the second-largest mango-producing state in India, and Chittoor district is the first-largest mango-producing district in Andhra Pradesh state. The dataset consists of 1344 images, which include samples of mango leaves affected by 13 different diseases and images of healthy mango leaves. The 13 different diseases represented in the dataset are Anthracnose, Apoderus, Gummosis, Leaf Webber, Mango Sooty, Multi diseased Leaf, Neutrician deficiency, Phoma Blight, Powdery Mildew, Red rust algae, Scale Insects, weaver ant, and Sooty Mould. These images were captured using a Realme phone camera, ensuring a practical and real-world scenario representation.

4 Methodology

This study contains 13 different diseases affecting mango leaves, along with samples of healthy leaves. These diseases include both bacterial and fungal infections, such as Anthracnose, Apoderus javanicus, Gummosis, Leaf Webber, Mango sooty blotch, Multi diseased leaf, Nutrician deficiency, Phoma blight, Powdery mildew, Red rust algae, Scale insect, Sooty mold, and Weaver ant. The study utilized a Convolutional Neural Network (CNN) approach using deep learning models to analyze the dataset. Three CNN models were tested in the study: GooGLeNet, ResNet-50, and EfficientNet. Figure 2 depicts the approach of the suggested system, which consists of six main components Data acquisition, feature extraction, classification, model evaluation, and prediction.

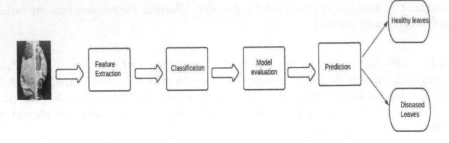

Fig. 2. Proposed Mango Leaf disease detection methodology

4.1   Feature Extraction

The preprocessed images are fed into the three pre-trained models, EfficientNet, ResNet-50, and GoogleNet, which extract features from the images.

4.2   Classification

The fused features are fed into a classification layer, which classifies the mango leaf images into different categories based on the type of disease they exhibit. The output of the classification layer is a probability distribution over the different categories.

4.3   Model Evaluation

The performance of the model is evaluated using different metrics, such as accuracy, precision, recall, and F1 score. The evaluation results are used to fine-tune the model and improve its accuracy.

4.4   Prediction

The trained model is used to predict the type of disease exhibited by new mango leaf images (Fig. 3).

The GooGLeNet model was trained on the dataset for 30 epochs with an initial learning rate of 0.001 and the SGDM optimizer, producing an accuracy rate of 20%. The number of epochs was then increased to 30, and the model was trained using the Adam optimizer, resulting in an accuracy rate of 61.54%. Further increasing the epochs to 50 improved the accuracy rate to 74.19%, again epochs are increased to 70 but there is no change in accuracy.

The ResNet-50 model was trained on the dataset, dividing it into training (70%) and validation (30%) datasets. The model was trained with an initial learning rate of 0.001 for 30 epochs, achieving an accuracy of 76.88%. The number of epochs was then increased to 70, and the model was trained using the

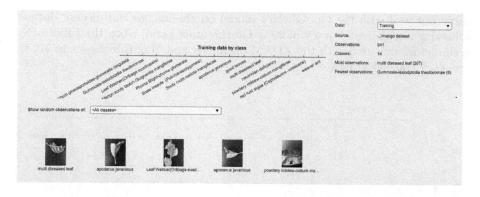

Fig. 3. A screenshot of the training Dataset

Adam optimizer, which increased the accuracy rate to 77.17%. Finally, the EfficientNet model was utilized, and it was trained for 30 and 70 epochs, achieving accuracy rates of 81.30% and 81.64% respectively (Figs. 4 and 5), Subsequently, with further training for 100 epochs, an impressive accuracy of 98.7% was achieved. The results suggest that the EfficientNet model performed the best among the three models tested in this study.

Table 1. Performance of training data based on a pre-trained model and epochs.

Model	Number of Epochs	Optimizer	Initial Learning Rate	Accuracy Rate
GooGLeNet	30	Adam	0.001	61.54%
GooGLeNet	70	Adam	0.001	74.19%
ResNet-50	30	Adam	0.001	76.88%
ResNet-50	70	Adam	0.001	77.17%
EfficientNet	30	Adam	0.001	81.30%
EfficientNet	70	Adam	0.001	81.64%

5   Results and Discussion

Based on the findings presented in Table 1, the training data was able to achieve varying levels of accuracy with different architectures. GooGLeNet achieved 74.19% accuracy with 70 epochs, ResNet-50 achieved 76.88% accuracy with 70 Epochs, and EfficientNet achieved 81.64% accuracy with 70 epochs. The accuracy of the models varied with the number of epochs used. The simulations showed that the training data set attained high accuracy in various instances, especially with the ADAM optimizer. However, the SGDM optimizer did not carry out well with the GooGLeNet model on the mango leaf disease dataset, achieving a training accuracy of 20%. On the other hand, when the EfficientNet architecture was used with the ADAM optimizer and Max Epochs set to 70, the accuracy reached 81.64%.

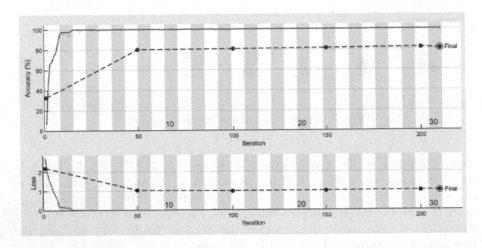

Fig. 4. EfficientNet with 30 epochs, Accuracy-81.30%

Figures 6 and 7 demonstrate the results of the training accuracy tests by utilizing ADAM and SGDM optimizers. Upon relating the two, it is evident that ADAM performed better.

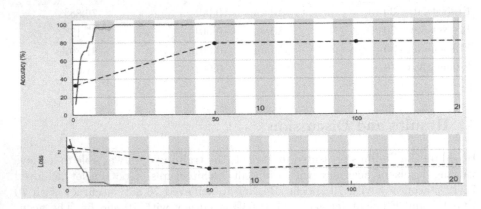

Fig. 5. EfficientNet with 70 epochs, Accuracy-81.64%

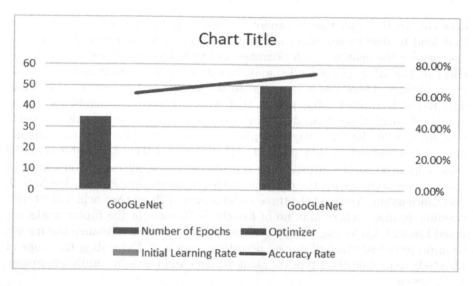

Fig. 6. Performance of GooGLeNet model with 35 and 50 Epochs, accuracy was same for 70 epochs

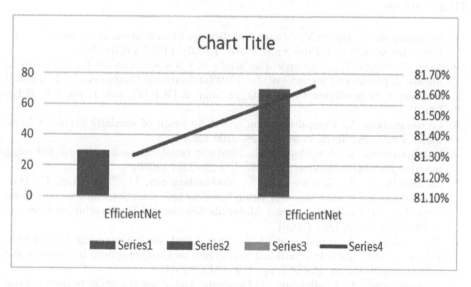

Fig. 7. Performance of EfficientNet model with 30 and 70 Epochs

6   Conclusion

It has become difficult for researchers to develop solutions that can be employed for the primary diagnosis of these disorders and to deal with the issues in a sustainable environment because of the complexity of the medicinal impacts of mango tree leaves. Nonetheless, it is commonly known that we have no control over the weather and that it cannot be modified. In contrast, several factors that lead to disease are under our control and can be anticipated early. Leaf diseases are the main reason for damage and yield loss, which has a devastating effect on the safety of the food supply. In extreme situations, it can even result in no harvest. Therefore, it is crucial to use mango leaf disease identification. The model is trained using photos that were obtained independently in our opinion, Convolutional Neural Network models are a great fit for this approach. Our main focus lay on using real photographs of the plants in the training data Better datasets would make it easier to reach the high performance of these applications. The investigation is performed by acquiring the outcomes on the training accuracy for the different models namely, GooGLeNet, ResNet-50, and EfficientNet. Amongst the three architectures, EfficientNet achieved 81.64% training accuracy where Max no of Epochs is 70 some of the future works are listed below. 1. Collecting a larger and more diverse dataset of mango leaf images to improve the accuracy of disease detection models. 2. Expanding the scope of the study to include other types of plant diseases and pests, in addition to mango leaf diseases

References

1. Arivazhagan, S., Ligi, S.V.: Mango leaf diseases identification using convolutional neural network. Int. J. Pure Appl. Math. **120**(6), 11067–11079 (2018)
2. Arya, S., Singh, R.: A comparative study of CNN and alexnet for detection of disease in potato and mango leaf. In: 2019 International Conference on Issues and Challenges in Intelligent Computing Techniques (ICICT), vol. 1, pp. 1–6. IEEE (2019)
3. Balasundaram, A.: Computer vision based detection of partially occluded faces. Int. J. Eng. Adv. Technol. **9**(3), 2188–2200 (2020)
4. Balasundaram, A., Ashokkumar, S.: Study of facial expression recognition using machine learning techniques. JCR **7**(8), 2429–2437 (2020)
5. Balasundaram, A., Ashokkumar, S., Kothandaraman, D., Sudarshan, E., Harshaverdhan, A., et al.: Computer vision based fatigue detection using facial parameters. In: IOP Conference Series: Materials Science and Engineering, vol. 981, p. 022005. IOP Publishing (2020)
6. Balasundaram, A., Chellappan, C.: Vision based motion tracking in real time videos. In: 2017 IEEE International Conference on Computational Intelligence and Computing Research (ICCIC), pp. 1–4. IEEE (2017)
7. Balasundaram, A., Chellappan, C.: Computer vision based system to detect abandoned objects. Int. J. Eng. Adv. Technol. **9**(1), 4000–4010 (2019)
8. Balasundaram, A., Chellappan, C.: An intelligent video analytics model for abnormal event detection in online surveillance video. J. Real-Time Image Proc. **17**(4), 915–930 (2020)
9. Balasundaram, A., Chellappan, C.: Vision based gesture recognition: a comprehensive study (2017). https://api.semanticscholar.org/CorpusID:31975735
10. Balasundaram, A., Kumar, S.A., Kumar, S.M.: Optical flow based object movement tracking. Int. J. Eng. Adv. Technol. **9**(1), 3913–3916 (2019)
11. Board, N.H.: National Horticulture Board (n.d.). Accessed 28 July 2023

12. Chen, J., Chen, J., Zhang, D., Sun, Y., Nanehkaran, Y.A.: Using deep transfer learning for image-based plant disease identification. Comput. Electron. Agric. **173**, 105393 (2020)

13. Emmanuel, T.O.: Plantvillage dataset. Kaggle. https://www.kaggle.com/datasets/emmarex/plantdisease. Accessed 2018

14. Ferentinos, K.P.: Deep learning models for plant disease detection and diagnosis. Comput. Electron. Agric. **145**, 311–318 (2018)

15. Fernández-Ponce, M.T., Lourdes, C., et al.: Potential use of mango leaves extracts obtained by high pressure technologies in cosmetic, pharmaceutics and food industries. Chem. Eng. Transac. **32** (2013)

16. Kothandaraman, D., Balasundaram, A., Korra, S., Sudarshan, E., Vijaykumar, B.: Enhancing dull images using discrete wavelet families and fuzzy. In: IOP Conference Series: Materials Science and Engineering, vol. 981, p. 022020. IOP Publishing (2020)

17. Kumar, S.M., Balasundaram, A., Sasikumar, A., Sathish Kumar, P.: Human emotion prediction-a detailed study. Eur. J. Mol. Clin. Med. **7**(4), 2020 (1991)

18. Merchant, M., Paradkar, V., Khanna, M., Gokhale, S.: Mango leaf deficiency detection using digital image processing and machine learning. In: 2018 3rd International Conference for Convergence in Technology (I2CT), pp. 1–3. IEEE (2018)

19. Mia, M.R., Roy, S., Das, S.K., Rahman, M.A.: Mango leaf disease recognition using neural network and support vector machine. Iran J. Comput. Sci. **3**, 185–193 (2020)

20. Pham, T.N., Van Tran, L., Dao, S.V.T.: Early disease classification of mango leaves using feed-forward neural network and hybrid metaheuristic feature selection. IEEE Access **8**, 189960–189973 (2020)

21. Prabu, M., et al.: A survey: benefits of mango leaves and techniques used for evaluvation of diseases affecting mango leaves. Inf. Technol. Ind. **9**(2), 507–512 (2021)

22. Prakash, B., Yerpude, A.: Identification of mango leaf disease and control prediction using image processing and neural network. Int. J. Sci. Res. Dev. (IJSRD) **3**, 794–799 (2015)

23. Prakash, O.: Diseases and disorders of mango and their management. In: Naqvi, S.A.M.H. (ed.) Diseases of Fruits and Vegetables Volume I: Diagnosis and Management, pp. 511–619. Springer, Dordrecht (2004). https://doi.org/10.1007/1-4020-2606-4_13

24. Sethupathy, J., Veni, S.: Opencv based disease identification of mango leaves. Int. J. Eng. Technol. (IJET) **8**(5) (2016)

25. Sharma, P., Berwal, Y.P.S., Ghai, W.: Performance analysis of deep learning CNN models for disease detection in plants using image segmentation. Inf. Process. Agric. **7**(4), 566–574 (2020)

26. Singh, U.P., Chouhan, S.S., Jain, S., Jain, S.: Multilayer convolution neural network for the classification of mango leaves infected by anthracnose disease. IEEE Access **7**, 43721–43729 (2019)

27. Sood, M., Singh, P.K., et al.: Hybrid system for detection and classification of plant disease using qualitative texture features analysis. Procedia Comput. Sci. **167**, 1056–1065 (2020)

28. Swetha, K., Venkataraman, V., Sadhana, G., Priyatharshini, R.: Hybrid approach for anthracnose detection using intensity and size features. In: 2016 IEEE Technological Innovations in ICT for Agriculture and Rural Development (TIAR), pp. 28–32. IEEE (2016)

29. Trang, K., TonThat, L., Thao, N.G.M., Thi, N.T.T.: Mango diseases identification by a deep residual network with contrast enhancement and transfer learning. In: 2019 IEEE Conference on Sustainable Utilization and Development in Engineering and Technologies (CSUDET), pp. 138–142. IEEE (2019)
30. Yamashita, R., Nishio, M., Do, R.K.G., Togashi, K.: Convolutional neural networks: an overview and application in radiology. Insights Imaging 9, 611–629 (2018)

RWNR: Radial Basis Feed Forward Neural Network Driven Semantically Inclined Strategy for Web 3.0 Compliant News Recommendation

Beulah Divya Kannan[1] and Gerard Deepak[2(✉)]

[1] Department of Computer Science and Engineering, National Institute of Technology, Tiruchirappalli, India
[2] Department of Computer Science and Engineering, Manipal Institute of Technology Bengaluru, Manipal Academy of Higher Education, Manipal, India
gerard.deepak.christuni@gmail.com

Abstract. In this paper, we propose a framework for socially responsible and semantically guided Web 3.0 compatible news recommendation that combines deep learning with strategic knowledge derivation. For this reason, RWNR model has been proposed. The RWNR model encompasses the radial basis feed forward neural network which is transformed into a feature controlled model by encompassing entities from the news API stack as features to classify the dataset and the classified dataset is subjected to the computation of strategic semantic similarity as well as step deviance at several phases with the reference tag pool which is generated from the user's topic of interest which is subjected to topic modelling utilizing Latent Dirichlet allocation (LDA) harvesting entities from wikidata as well as making it socially aware by encompassing the twitter API. Hybridization of entities from the news API stack is done from several knowledge derivation schemes into the model. The normalized information distance (NID) and tag index with the harmony search optimization algorithm is metaheuristic in nature which optimizes the solution that makes the proposed model not only rich in knowledge but also rich in semantics and optimization. This has resulted in best-in-class model with average recall, F-measure, False discovery rate (FDR), accuracy percentages of 96.09%, 91.19%, 96.64% and 96.63% respectively with the lowest FDR of 0.04.

Keywords: Latent Dirichlet allocation · News Recommendation · Web Crawler · Radial Basis Function Neural Network (RBFNN)

1 Introduction

In the modern era, news recommendation systems are essential to combat information overload, offer individualized news experiences, expose people to other viewpoints, help people find new topics, and guarantee the timeliness and relevancy of news consumption. They increase user pleasure, save up time, and promote an informed and active society. People are frequently overwhelmed by the sheer amount of information available due

© The Author(s), under exclusive license to Springer Nature Switzerland AG 2023
S. Tiwari et al. (Eds.): AI4S 2023, CCIS 1907, pp. 145–157, 2023.
https://doi.org/10.1007/978-3-031-47997-7_11

to the rapid expansion of digital content and the multiplicity of news sources. Systems for recommending news articles aid in resolving this issue by curating and filtering news material according to user preferences. By presenting material that is pertinent and catered to their individual likes, news recommendation systems give users the opportunity to have a personalized news experience. These systems can suggest news articles, videos, and other content based on user behavior, browsing history, and demographic data.

Overall, this paper deals with how news recommendation system improves news consumption by utilizing technology and machine learning, which helps users save time, gain new knowledge, and develop a more comprehensive perspective of the world.

Contribution: The RWNR encompasses the radial basis feed forward neural network which is transformed into a feature-controlled model by hybridizing four news API stacks namely the newsAPI.org, Bloomberg API, Guardian Platform and BBC platform which are used as standard feature selectors used for classifying the dataset. The user's topic of interest is preprocessed and enriched by applying Latent Dirichlet Allocation (LDA) which is a strategic topic model, wikidata API which is used to harvest community-contributive and community-verified hyperlinked entities and the social elements are further annotated to assess and encompass social awareness into the proposed RWNR model. Strong relevance computation mechanisms are ensured in the model by the normalized information distance (NID) and tag index with differential thresholds and empirically determined step deviation measurements, and the harmony search algorithm optimizes user search and produces the most ideal entities that are most pertinent to the topic entities without compromising on the diversity as well as the significance of the results. Overall, the accuracy, recall, precision, f-measure and FDR are calculated for all the models, the proposed architecture out performs the baselined models.

Organization An overview of the concept is provided in the paper's introduction. The remainder of the structure of the paper is outlined below: Work pertaining to the area of study and experimentation is shown in Sect. 2. The proposed design for the system is shown in Sect. 3. Implementation is covered in Sect. 4. Experimental findings are presented in Sect. 5. The conclusion and references are in Sect. 6.

2 Related Works

A news recommendation algorithm with customized attention was put up by Chuhan et al., [1] for the user to find the news of their specific interest and to minimize the overload of information. This work suggests a customized attention network that makes use of the user ID embedding to provide a request word- and news-level attention-vectors. A news recommending system using a combined deep learning approach was suggested by Gabriel et al., [2]. The paper offers a novel method that enables weighing potential compromises, such as accuracy vs. uniqueness for diverse information sources., according to the particular requirements of the recommendation system. Two publicly available news datasets were used to evaluate the studies. In order to create accurate representations of users and news, Fangzhao et al., [3] suggests a neural news recommendation strategy that makes use of many forms of news data. The strategy employs a user encoder that learns user representations based on how users browse the news and a news encoder that assembles cohesive news representations from topic categories, titles, and bodies. Experimental findings show that this strategy significantly improves news recommendation on a real-world dataset. Hu et al., [4] offers a system for news recommendation that models user, news, and latent topic interactions on a heterogeneous graph. Topic information is incorporated into the model to account for the lack of user-item interactions and to apprehend user interests. Dhruv et al., [5] proposes a 3D convolutional neural network for news recommendation systems. It takes into account both the news story's context and the order in which the papers were read by the users. On the emphasis of topic-aware news representations, Chuhan et al., [6] suggests a neural network-based news recommendation system. In the encoder block, CNN networks are applied in order to get the context of the title and special attention-based mechanisms are used to get the important words from the title for the recommendation system. Various deep learning mechanisms have been incorporated for news recommendation systems as seen in [7, 8] and [9]. For instance, Jianxun et al., [9] has suggested a deep learning-based model that makes use of CNN based inception module and attentive fusion method is used in this paper. The paper calls for mechanical feature engineering, which requires extensive subject expertise and time to create. Additionally, these techniques are unable to capture word contexts and ordering in recommendation systems, which is crucial for news learning. In [7, 10, 11], and [12] several news recommendation strategies encompassing several paradigms have been depicted but all of these strategies do not use collective intelligence and lack knowledge driven semantics. Chhatwal et al., [13] and Srivastava et al., [14] have proposed semantically inclined models for multisource hybridization of recommendation of news with an emphasis on personalization and emotional Forecasting from News Articles for Intermediary Readers. Several semantically inclined models in [15, 16] and [17] for encompassing ontologies and knowledge graphs have been proposed for similar frameworks.

3   Proposed Architecture

Figure 1 shows the suggested system design of the knowledge-centric framework for news recommendation. Preprocessing of the user queries entails stop word removal, lemmatization, tokenization and named entity recognition (NER). For this, word sense disambiguation (WSD) is performed. So once the original topic words which are individual words in its base form is obtained, it is subjected to topic modelling on application of the Latent Dirichlet Allocation (LDA) model by using the present structure of the World Wide Web (WWW) as the reference corpora. Once the user topics are generated and modelled by adding topics, The LDA (Latent Dirichlet Algorithm) is encompassed for topic enrichment of topic words and subjected to two new API's namely the Twitter API and Wikidata API. Wikidata is knowledge stored which comprises of community-based hyperlinked instances of knowledge. Twitter API is an application programming interface which has access to hashtags as well as the trending topics on Twitter. So, the entities which come out of the Wikidata and Twitter APIs are further subjected to a trending tag pool where the actual trending tags from twitter are preprocessed and the trending tag pool is formalized. From the dataset end, the dataset is subjected to

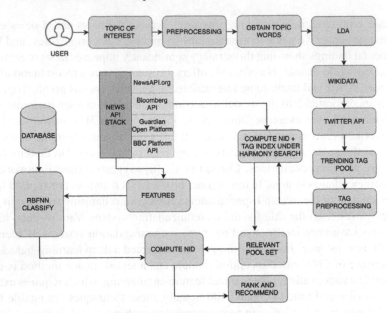

Fig. 1. Proposed system architecture

classification of Radial Basis Function Neural Network (RBFNN) in which features are selected from large scale knowledge stack namely the News API stack that is News-API.org, Bloomberg news API, Guardian Open Platform and BBC platform API. These are the four distinct news API platforms which are used to identify news that is trending to the current news and moreover four different news APIs are used because of coherence at which the news is trending, validated and also because news is at different time zones and different geographies. These APIs are then formalized. Therefore, from the different News API stacks, the enrichment of the news would be increased. From this, the features are extracted and these features are subjected to radial basis function neural network (RBFNN) classifier randomly to classify the instances that come out of the dataset which are further subjected to computation of normalized information distance (NID) with the relevant pool. This is in turn derived from computing the normalized information distance and the tagged index and harmony search algorithm between the tags which are preprocessed that come out from the trending tag pool and on top of that entities are discovered from the news API. Entities and news API are large scale in nature and the tags which come out from Wikidata API and twitter API are also large scale in nature, there is a need for an optimized algorithm like a harmony search algorithm to yield optimal results for which the initial tendency set is obtained by applying the normalized information distance (NID) and TAG index. The optimal solution set is then yielded by computing the harmony search with the empirical threshold of 0.75 for normalized information distance and 0.25 for Tag index by keeping the empirical threshold very strong as there are large number of instances going into the pipeline. To yield the relevant pool set, the relevant pool entity set is subjected to computation of normalized information distance (NID) to that of the entities which come out of the radial basis function neural network (RBFNN) classifier. At this end, the NID is set to a threshold of only 0.5 relevance and most optimal entities are already discovered at one end and at the other end, the entities are classified using the radial basis function neural network (RBFNN) classifier. Therefore, due to this dataset, the NID is set to a threshold 0.5 owing to the strength of Normalized information distance and finally the entities which come out of this pipeline are ranked and suggested in ascending order of NID and the topics comprising of the user's topic of interest is yielded and all the trending news with these hashtags on twitter and with these keywords on the news API stack and the dataset is recommended to the end-user. This process continues until there are no more logged user clicks.

To examine time series or text-based sequential data, recurrent neural networks were developed. RNNs have feedback connections that enable information to flow both forward and backward over time, in contrast to feedforward neural networks. In order to capture and process data from earlier steps in the sequence, RNNs must be able to maintain an internal memory, also known as a hidden state. The network uses this hidden state, which is updated at each time step, as context to comprehend the current input in light of previous inputs. RNNs can handle sequences of any length since they operate by recursively applying the same set of weights and biases across each time step. Each step's output from an RNN can be used for prediction or transferred to the step after it in the sequence.

The Long Short-Term Memory (LSTM) network is a well-liked RNN version that eliminates the problem of vanishing gradient issues by including specialized memory cells that can store data for extended periods of time. LSTMs have been extensively employed in a range of activities, such as speech recognition, machine translation, and language modelling. They are particularly good at capturing long-range dependencies in sequences. A forget gate, output gate and input gate are the three integral parts of an LSTM cell. When no longer required, certain types of state information held in an RNN's internal state can be erased, according to the Forget Gate. The Input side gate makes the decision of whether fresh data that may be beneficial to the model is added to long-term memory or cell state. Which portion of the input gate should be displayed as the output in that specific situation is specified by the output gate. The equations for Long short-term memory are given below from Eq. (1) to Eq. (6).

$$a_t = \sigma_g(H_f \times x_t + G_f \times h_{t-1} + b_f) \tag{1}$$

$$i_t = \sigma_g(H_i \times x_t + G_i \times h_{t-1} + b_i) \tag{2}$$

$$o_t - \upsilon_g(H_o \times x_t + G_o \times h_{t-1} + b_o) \tag{3}$$

$$c'_t = \sigma_c(H_c \times x_t + G_c \times h_{t-1} + b_c) \tag{4}$$

$$c_t = f_t \cdot c_{t-1} + i_t \cdot c't \tag{5}$$

$$b_t = o_t \cdot \sigma_c(c_t) \tag{6}$$

where c_t is cell state, b_t is hidden state, a_t is forget gate, i_t is input gate, o_t is output gate and $H_f, H_i, H_o, H_c, G_f, G_i, G_o$ and G_c are the respective weights.

This model applies a sigmoid function to the input word x(t) and the prior hidden state (short-term memory), and returns a value between 0 and 1. This step is represented by Eq. (1). It will need to output a zero vector in order to erase the prior memory or state, and when multiplied by the previous cell state, the result will be 0. This denotes the resetting of the memory. The sigmoid and tanh functions are utilized at the input gate, shown in Eq. (2), to add the present input x(t) into the cell state (long term memory). The weighed sum of h(t-1) and x(t) is now fed onto a Sigmoid function at the output gate Eq. (3), and the tanh function is applied to the cell state. Equation (4) depicts the candidate cell state. Finally, we multiply the value in the output gate by the result of the tanh function in Eq. (6) to obtain our hidden state h(t). Equation (5) can be used to determine the cell state or long-term memory, which is c(t), and the output, or short-term memory, which is h(t).

RBFNN: Radial Basis Function Neural Network is a fast and intuitively used Machine Learning algorithm. Both problems involving regression and classification are resolved using this approach. It is mainly used in pattern recognition tasks. Radial basis functions are primarily used by RBFNN as activation functions. Three layers make up the radial Basis function Neural Network, an artificial neural network. The primary layer is linear and distributes those signals that are used as an input in this neural network. The second layer is nonlinear and makes use of Gaussian functions. The third layer linearly combines the output of the gaussian functions.

The equation of a basic radial basis function neuron is given below in Eq. (7):

$$\varnothing(x) = exp\left(-\frac{||x - o||^2}{2\sigma^2}\right) \tag{7}$$

x = input to the neuron
o = center of the neuron
σ = the spread of the gaussian function
$||.||$ = represents the Euclidean distance

The weighted sum of the activation functions of the input to the following radial basis function neural network constitutes the output of the RBF neuron which is shown below in Eq. (8)

$$y = \sum_{i=1}^{n} w_i \varnothing(x_i) \tag{8}$$

y = output of the RBF neuron
w_i = weight associated with each input
$\varnothing(x_i)$ = activation of each neuron corresponding to each input xi

This algorithm overcomes various disadvantages that traditional gradient problems face like the no of epochs to be trained on, learning rate, stopping criterion, local minima etc. Moreover, due to its shorter training time and generalizing capability, radial basis function neural networks are employed in present time applications.

Latent Dirichlet Allocation: Latent Dirchlet Allocation model is one of the most popular topic modelling methods in which it annotates each document based on the topics predicted by the topic modelling method and gives out optimized search results. Another statistical model utilized in problems involving natural language processing is LDA. The basic objective of the LDA model is to produce topics from a varied range of documents depending on the words' occurrence. It is also used to find mixtures of topic for a given document. For example, to find a particular news in our case, a collection of news documents is collected and created. Each document collected represents a news article. Data cleaning is done on the collected documents which is usually done in various NLP tasks. The first step in data cleaning is tokenizing in which each document is converted to its atomic elements. The second phase is stop word removal, which involves removing words that have no meaning. The third step is stemming where words are merged that are equivalent in meaning. The LDA model assigns a random topic to each word in the corpus of the documents that has been collected. It focuses on maximizing the separability of known categories and the particular topics with respect to the news articles are recommended. In general, A statistical technique called latent Dirichlet allocation can be used to find hidden themes in a collection of documents. It makes the assumption that documents are collections of topics and that words are produced depending on those topics. A means to comprehend the underlying thematic structure of a document collection is made possible by LDA by inferring the hidden topic distributions and word distributions.

Harmony Search Algorithm: The Harmony Search (HS) method is an improvisation-inspired optimization algorithm. To identify the best answer to an optimization problem, it imitates how musicians create harmonies. The program keeps track of a population of potential solutions in a harmony memory. By merging parts from the memory and adding random changes through pitch modification, a new harmony is created in each repetition. The objective function of the problem is used to evaluate the new harmony, and if it performs better than a harmony in memory, that harmony is updated. Until a termination condition, such as exceeding a maximum number of iterations or convergence criteria, is satisfied, this iterative process continues. In a nutshell the Harmony Search algorithm draws inspiration from musical improvisation to best solve a particular problem. It keeps track of potential answers in a memory, combines ideas from the memory to form fresh harmonies, and modifies the pitch to add randomization. The algorithm seeks to identify the best solution that maximizes or reduces the objective function of the optimization problem through an iterative process of creating, assessing, and modifying harmonies.

4 Implementation

The implementation is carried out using Python. The TensorFlow framework is used for building the radial basis function neural network (RBFNN) and configuring it as a feature-controlled configuration. Certain code level modelling was done with RBFNN so it can fit the features from API. Standard APIs were directly accessed based on topic of relevance and python's NLTK framework was used for preforming all the NLP operations like preprocessing. Topic modelling was again fit using LDA and the web as a reference corpus. Wikidata was accessed by standard API and the twitter API was also accessed into the framework.

The data set which was used as a single large constituent dataset was derived from three independent datasets namely the NewsREC dataset by Gabriel Machado et al.,

[18]. The second dataset used is the Adressa Dataset [19]. The third dataset is USA News Dataset by Vinayak Shanawad [20]. These datasets were made sure that they were integrated into a single large platform by crawling them, by putting them into separate ESP files, preprocessing the individual tuples and apart from that annotating them by adding categories by crawling the current structure of the world wide web and if they were not categorical in nature, the keywords were extracted and these keywords were structurally fit as annotations and they were further annotated such that many labels and annotations were generalized and based on these annotations, it was rearranged and reprioritized into a single large news dataset and this was made use of in order to conduct the experiment.

Experimentations were conducted for 7281 novel topics which were single word as well as multi-word not more than ten words as topic of interest which is considered as queries and based on these topics, experimentations were conducted and F-measure, accuracy, precision, recall were formalized using the standard formulations and False discovery rate was also formalized using a standard formulation.

5   Results and Performance Evaluation

Accuracy, F-measure, Recall, Precision Percentages, and False Discovery Rate (FDR) are considered as suitable measures to measure the performance of the suggested RWNR architecture. F-measure, Recall, Precision, and Accuracy calculate and quantify the importance of the results, whereas the false discovery rate calculates the number of false positives suggested by the proposed architecture, which denotes the error rate. It is evident from Table 1 that the suggested RWNR architecture produces the greatest precision average percentage (96.09%), highest recall average percentage (97.19%), highest accuracy average percentage (96.64%), highest F-measure average percentage (96.63), and lowest FDR (0.04). The proposed RWNR structure is baselined with four different models, namely the NNRL [10], OBNR [11], PNRCT [7], and DKNR [12] frameworks, which are also prospective news recommendation models and share a similar set of assumptions and goal. This allows for an evaluation of the proposed RWNR's performance and a comparison of the suggested model with other models.

The NNRL architecture produces overall average precision of 88.27%, overall average recall of 90.07%, overall average accuracy of 89.17%, and overall average F-measure of 89.16% with FDR of 0.12. The OBNR model provides average values for precision, accuracy, F-measure, and accuracy with an FDR of 0.10 of 90.07%, 93.02%, 91.54%, and 91.52%, respectively. The PNRCT model provides average values for precision, recall, accuracy, and F-measure with an FDR of 0.07, 93.08% average for precision, 94.09% average for recall, and 93.58% average for accuracy and 95.58% for F-measure. The DKNR framework provides average values for accuracy, precision, recall, and F-measure of 94.24%, 93.39%, 97.16%, and 94.23%, respectively, with an FDR of 0.07. Because it is a deep learning framework that is motivated by semantics, the suggested RWNR framework produces the greatest average recall, precision, accuracy, F-measure percentages, and lowest False discovery rate (FDR) results. The radial basis function neural network (RBFNN) model which is a neural network framework is fed with features and it is configured in such a way that it accepts features and the features are accessed from the news API stack which comprises of four distinct news APIs which are from different geographies, different vantage points but collage similar news for a similar timeline namely the NewsAPI.org, Guardian Open Platform, Bloomberg API and BBC platform API.

Table 1. Performance Evaluation of the Proposed RWNR Compared to Other architectures

Model	Average Precision percentage	Average Recall percentage	Average Accuracy percentage	Average F-Measure percentage	FDR
NNRL [10]	88.27	90.07	89.17	89.16	0.12
OBNR [11]	90.07	93.02	91.54	91.52	0.10
PNRCT [7]	93.08	94.09	93.58	93.58	0.07
DKNR [12]	93.39	95.09	94.24	**94.23**	0.07
Proposed RWNR	96.09	97.19	96.64	**96.63**	0.04

Furthermore, the framework obtains the query of interest and increases the density of auxiliary knowledge on the topic by putting it through the Latent Dirichlet Allocation (LDA) model, a topic modelling framework that automatically finds the topics that are pertinent but as of yet hidden to the topic of interest of the user by using the current web structure as the reference corpora. Further, the Wikidata encompasses community-contributive, community-verified hyperlinked entities from several heterogenous sources and thereby increasing the density of auxiliary knowledge. Furthermore, the twitter API discovers hashtags, words and terms which are socially popular, socially accepted and socially trending. From the Wikidata and twitter APIs, the entities are further used in order to increase the overall auxiliary density which not only yields a community coherence but also yields a social coherence as well. Most importantly, the encompassment of normalized information distance and Tag index with differential thresholds and step deviance measures which are empirically decided and the encompassment of harmony search algorithm for optimization ensures the proposed RWNR framework which is an encompassment of strategic deep learning model namely the radial basis function neural network with feed forward network is feature constructive in order to avoid deviance and collates knowledge entities as well as terms of social importance and relevance from heterogenous sources along with LDA driven topic modelling makes this model quite unique. Strong relevance computation mechanisms like the normalized information distance, tag index at different stages and phases fit into the framework. Also, the harmonic search algorithm that is a metaheuristic optimization algorithm plays a vital role in this framework. This makes it so that the suggested RWNR model beats all baselined models and produces the best F-measure, precision, recall, and accuracy. It also assures that the FDR is as low as possible.

The NNRL model does not function as anticipated because of the stated reasons. Although neural news recommendation system with long and short representations are fed into the model, the GRU network is encompassed. The short-term and long-term representations are shallow knowledge models. Feeding them into a strong deep learning model requires high density knowledge makes it a misfit between the strong deep learning models that requires GRU and the knowledge becomes quite minimalistic when compared. Hence, owing to this demarcation, the NNRL model results in loss of useful knowledge. Thereby the classification yielded for the news recommendation model is not very relevant when measured with the proposed model. Thus, NNRL model doesn't perform as anticipated.

The OBNR model lacks due to the following reasons. Although the OBNR model is an ontology-based model, the ontologies along with the most documented frequency work well for a specific domain but news is generalized. Several general-purpose ontologies or specialized ontologies has to be fed in for which it is not realistically available. Although if it yields a higher measure of recall, accuracy, precision and F-measure, it might result in a better accuracy at times but there is always a shallow knowledge embedding because the ontologies are light weight models. Stronger ontologies with TF-IDF with lighter or absence of semantics is believed to be a mismatch. Here, knowledge is represented at a perfect knowledge model in terms of ontologies. TF-IDF is incorporated to signify knowledge but the knowledge attenuation and semantics between the instances of knowledge is definitely absent. Hence, the OBNR model also doesn't implement well as anticipated.

The reason why DKNR model doesn't perform well when compared to the proposed models is because the DKNR framework which is deep knowledge of a network. Convolution neural network and knowledge graphs are incorporated in this framework. Semantics are also fairly strong for this framework but the amount of knowledge which is incorporated into the convolutional neural network is insufficient. Word entity aligned knowledge alone is represented. It is not discovered from the world wide web so knowledge infusion into the model is not very efficacious and semantic reasoning is also not very strong. There is a need for optimization model into this framework which is absent. When optimization model is fit into the DKNR model, high levels of underfitted results would be obtained. The DKNR model consequently doesn't enact as predicted. In comparison to the suggested architecture, the PNRCT model similarly exhibits poor performance. The reason is because the PNRCT which is personalized News recommendation with context trees are light weight and best in class models. But in reality, there are no context trees that can be obtained from the structural metadata. In the present era, this model could perform well if there a perfect Web 3.0 framework is built. But at this point, the context trees which have been derived from the browsing behavior of the user would definitely yield to underfitting and there is no strong learning into the framework and semantics is weak and auxiliary knowledge is also quite weak and shallow. Therefore, the PNRCT model doesn't do well.

Owing to all these reasons, the suggested RWNR architecture is a deep learning model which mandatorily incorporates knowledge from the news API stack from four different heterogenous sources as features to classify the dataset. Also, outgrowing the topic of interest using topic modelling in terms of LDA, wikidata and Twitter APIs which ensures high degree of knowledge is added and strong relevance computation mechanisms in terms of normalized information distance and tag index with empirically decided thresholds and metaheuristic optimization using harmony search algorithm outperforms all the baselined models as it provides strong optimization with robust computational techniques for semantic relevance along with high density auxiliary knowledge collated from several heterogenous resources which makes the RWNR model the best model.

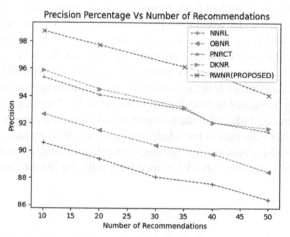

Fig. 2. Precision percentage Vs No. of Recommendations

Precision percentage Vs Number of Recommendations line graph curve is depicted in Fig. 2. It is evident that the top spot in the hierarchy is held by the proposed RWNR model. The lowest place is occupied by the NNRL model. The second place is occupied by the DKNR model. The penultimate position is occupied by the OBNR model in the hierarchy. The mid position is occupied by the PNRCT model in the line graph curve. The RWNR model takes the first position because the framework encompasses of LDA, Wikidata API which is community-contributive, community-verified and contains hyperlinked entities from heterogenous resources. It also makes use of the Twitter API where high degree of knowledge is added to the framework and strong computation mechanisms are used which makes secure the first position. The disadvantage of the DKNR model is that semantic reasoning and knowledge infusion into this model is not as strong compared to the other models which makes it a misfit. The drawback for the PNRCT model is that the context trees that have been derived from browsing behavior of the user leads to underfitting as there is no strong learning into the framework. The drawback for the OBNR model is due to the shallow knowledge embeddings in the framework. The NNRL performs the worst due to the presence of long term and short-term representations that are shallow knowledge model. Feeding this to the network requires a strong deep learning model, which this model doesn't satisfy and leads to a misfit.

6 Conclusion

This paper suggests a strategic framework for news recommendation compliant with web 3.0. The proposed RWNR framework is driven by user's topic of choice which is enriched by subjecting it to topic modelling in terms of Latent Dirchelet Allocation (LDA) which is further enriched from community-contributive and community-verified entities harvested from the wikidata API and furthermore the socially aware entities in terms of tags which are trending in twitter are also incorporated in order to ensure and contextualize the user required topic terms in terms of the current socially relevant news embeddings. The radial basis function neural network (RBFNN) is employed for classifying the dataset which is in turn controlled by features which is encompassed from a news stack which consists of four distinct top news APIs. The normalized information distance (NID) and tag index with the harmony search optimization algorithm makes the proposed model not only rich in knowledge but also rich in semantics and optimization which has achieved highest average accuracy, recall, precision, f-measure percentages of 96.09%, 97.19%, 96.64%, 96.63% respectively with the lowest False discovery rate (FDR) of 0.04.

References

1. Wu, C., Wu, F., An, M., Huang, J., Huang, Y., Xie, X.: NPA: neural news recommendation with personalized attention. In: Proceedings of the 25th ACM SIGKDD International Conference on Knowledge Discovery & Data Mining, pp. 2576–2584 (2019)
2. Moreira, G.D.S.P., Jannach, D., Cunha, A.M.D.: Contextual hybrid session-based news recommendation with recurrent neural networks. IEEE Access 7, 169185–169203 (2019). https://doi.org/10.1109/ACCESS.2019.2954957
3. Wu, C., Wu, F., An, M., Huang, J., Huang, Y., Xie, X.: Neural news recommendation with attentive multi-view learning (2019). arXiv preprint arXiv:1907.05576
4. Hu, L., Li, C., Shi, C., Yang, C., Shao, C.: Graph neural news recommendation with long-term and short-term interest modeling. Inf. Process. Manage. 57(2), 102142 (2020)
5. Khattar, D., Kumar, V., Varma, V., Gupta, M.: Weave&rec: a word embedding based 3-D convolutional network for news recommendation. In: Proceedings of the 27th ACM International Conference on Information and Knowledge Management, pp. 1855–1858 (2018)
6. Wu, C., Wu, F., An, M., Huang, Y., Xie, X.: Neural news recommendation with topic-aware news representation. In: Proceedings of the 57th Annual Meeting of the Association for Computational Linguistics, pp. 1154–1159 (2019)
7. Wang, H., Zhang, F., Xie, X., Guo, M.: DKN: deep knowledge-aware network for news recommendation. In: Proceedings of the 2018 World Wide Web Conference, pp. 1835–1844 (2018)
8. Okura, S., Tagami, Y., Ono, S., Tajima, A.: Embedding-based news recommendation for millions of users. In: Proceedings of the 23rd ACM SIGKDD International Conference on Knowledge Discovery and Data Mining, pp. 1933–1942 (2017)
9. Lian, J., Zhang, F., Xie, X., Sun, G.: Towards better representation learning for personalized news recommendation: a multi-channel deep fusion approach. In: IJCAI, pp. 3805–3811 (2018)
10. An, M., Wu, F., Wu, C., Zhang, K., Liu, Z., Xie, X.: Neural news recommendation with long-and short-term user representations. In: Proceedings of the 57th Annual Meeting of the Association for Computational Linguistics, pp. 336–345 (2019)

11. IJntema, W., Goossen, F., Frasincar, F., Hogenboom, F.: Ontology-based news recommendation. In: Proceedings of the 2010 EDBT/ICDT Workshops, pp. 1–6 (2010)
12. Garcin, F., Dimitrakakis, C., Faltings, B.: Personalized news recommendation with context trees. In: Proceedings of the 7th ACM Conference on Recommender Systems, pp. 105–112 (2013)
13. Chhatwal, G.S., Deepak, G., Sheeba Priyadarshini, J., Santhanavijayan, A.: SemKnowNews: a semantically inclined knowledge driven approach for multi-source aggregation and recommendation of news with a focus on personalization. In: Patel, K.K., Santosh, K.C., Patel, A., Ghosh, A. (eds.) Soft Computing and Its Engineering Applications. icSoftComp 2022. Communications in Computer and Information Science, vol. 1788, pp. 263–274. Springer, Cham (2022). https://doi.org/10.1007/978-3-031-27609-5_21
14. Srivastava, R.A., Deepak, G.: PIREN: prediction of intermediary readers' emotion from news-articles. In: Shukla, S., Unal, A., Varghese Kureethara, J., Mishra, D.K., Han, D.S. (eds.) Data Science and Security. LNNS, vol. 290, pp. 122–130. Springer, Singapore (2021). https://doi.org/10.1007/978-981-16-4486-3_13
15. Ortiz-Rodriguez, F., Tiwari, S., Panchal, R., Medina-Quintero, J.M., Barrera, R.: MEXIN: multidialectal ontology supporting NLP approach to improve government electronic communication with the Mexican Ethnic Groups. In: DG. O 2022: The 23rd Annual International Conference on Digital Government Research, pp. 461–463 (2022)
16. Yadav, S., Powers, M., Vakaj, E., Tiwari, S., Ortiz-Rodriguez, F., Martinez-Rodriguez, J.L.: Semantic based carbon footprint of food supply chain management. In: Proceedings of the 24th Annual International Conference on Digital Government Research, pp. 657–659 (2023)
17. Vakaj, E., Tiwari, S., Mihindukulasooriya, N., Ortiz-Rodríguez, F., Mcgranaghan, R.: NLP4KGC: natural language processing for knowledge graph construction. In: Companion Proceedings of the ACM Web Conference 2023, p. 1111 (2023)
18. Lunardi, G.M., de Oliveira, J.P.M.: NewsREC dataset: news recommendation and diversification. In: Elsevier Applied Soft Computing (1.0), vol. 97. Zenodo (2021). https://doi.org/10.5281/zenodo.4604008
19. Gulla, J.A., Zhang, L., Liu, P., Özgöbek, Ö., Su, X.: The Adressa dataset for news recommendation. In: Proceedings of the International Conference on Web Intelligence, pp. 1042–1048. ACM (2017)
20. https://www.kaggle.com/vinayakshanawad/us-news-dataset (2021)

WDNRegClass - A Hybrid ANN + Bayesian Learning Model to Reduce Temporal Predictive In-Variance Towards Mitigation of WDN Revenue Losses

C. Pandian[1]([✉])[iD] and P. J. A. Alphonse[2][iD]

[1] Research Scholar, Computer Applications Department,
National Institute of Technology, Tiruchirappalli, Tamilnadu, India
pndn.dnl@gmail.com
[2] Professor, Computer Applications Department,
National Institute of Technology, Tiruchirappalli, Tamilnadu, India
alphonse@nitt.edu

Abstract. WDN (Water Distribution Network) leakages are observed to globally monetize around 39 billion USD annually, leading to greater loss in resource and revenue, infrastructural degradation, and related adversarial impacts. Appropriate monitoring, mitigation and management of leakage localization and isolation can bring out relative benefits in reducing the loss. So, our proposed work is on data-driven leakage identification and pressure-based feature analytics for localization upon the widely adopted BattLeDIM 2020 dataset for timely reduction of resource and revenue losses incurred. The proposed work is trained on pairwise pressure differential sensor data with self-supervised regression (Bayesian Learning) to gather proximity of the sensing signals. At the same time, abrupt leakages are identified through sigmoidal classification of local sensors to generate alarms in time. As a result, experiments show that the accuracy of the model is improved around 1.5–3% over existing systems.

Keywords: Sustainable Development Goals · Water Leak Prediction · Pressure variation exploration · Bayesian Learning · Regression and Classification Models

1 Introduction

Water losses are the third major non-renewable resource loss stated in SDG where leakage in water distribution networks (WDNs) [15] are being a major cause. The difference among the water supplied in WDNs [4] and the amount that's billed exceeding 120 billion cubic meters annually is used to estimate globally water loss., with 39 billion USD associated with it [26]). The main difference between a well-maintained WDN and poorly maintained WDNs account to background leakages and pipe bursts cause 3–7% and 35% losses respectively which is extremely higher on the second case [24]. Also, around 50% is lost for systems in developing countries [21] caused by pressure drops from pipe bursts and background leakages that continue to be unsolved long time.

The management of non-revenue water resources (NRWs) is anticipated to produce a number of advantages, including increased operational profits, reduced energy requirements, improved customer satisfaction, increased water services overall, and less environmental degradation [4]. With a strong body of literature dating back more than 50 years, finding of physical leaks has higher priority among the common utility analysis techniques used for NRW management (Puust et al. 2010). Due to the likelihood that WDNs with high pressure pipes can branch out into a large network system requiring stringent monitoring and diagnostics, as shown in , the detection of a leak in a WDN is also treated as a distinct convex set problem Fig. 1.

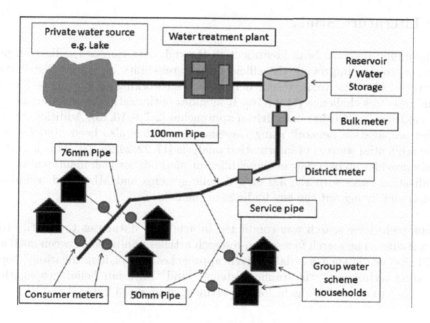

Fig. 1. Leak Scenario Overview [20]

Additionally, the challenge is made more difficult by changes in water demands, seasonal trends, and measurement noise [12–14] We categorize leakage detection assessment and control methods into three categories: leak detection (LD), leak isolation (LI), and leak localization (LL), in accordance with early studies that concentrate on identifying the potential factors that could cause leakages. Methods for locating or identifying leaks typically combine optimization techniques with hydraulic model parameters, [6,11,13,28], or data-driven approaches using pressure of flow data with high temporal resolution (i.e., subhourly), [9], CNN, [16], Bayes classifiers, and long short-term memory (LSTM) networks. Due to fewer parametric restrictions or higher bandwidth utilization, only a very small number of these systems have come close to saturation [3,14]. Leakage Detection and Isolation Methods are engaged in a battle. In order to meet the demands of prompt detection and location accuracy, an international competition called the Battle of the Leakage Detection and Isolation Methods is organized [20]. In this paper, we present a pressure-based LD approach that makes use of supervised classification algorithms and pairwise linear regression to improve leakage prediction, thereby reducing resource and revenue losses brought on by WDN [4,6,7].The remainder of the essay is structured as follows: The literature is expanded upon in Sect. 2, the LD approach and its modules are proposed in Sect. 3, and the empirical findings are discussed and reported in Sect. 4. Finally, in Sect. 5, we provide our conclusions and recommendations for additional research.

2   Literature Study

In hydraulic systems, Leak identification through detection or localization play a major role in improving the efficiency and reliability of Water Distribution Networks. Over the years, researchers have been working on various techniques to address this challenge [11,14,16]. It includes optimization, hydraulic parameter modelling, and other data-driven approaches [2,7,9,10,13]. Additionally, the belief propagation network using Bayesian theory has also been used for multiple sequential sources of information analysis [17,22,24,25,27]. This literature study provides an overview of these different methods for leak identification and localization tasks with the aid of hydraulic systems and ML based techniques [3,4,7] and listing out the key findings in the field.

A comprehensive search was conducted in academic databases to identify relevant studies. The search focused on research articles published between 2000 and 2023. Keywords such as "data-driven approaches," "leak identification," "optimization techniques," "hydraulic systems," and "Bayesian belief propagation" were used to ensure the inclusion of pertinent studies [13,14].

2.1   Leak Identification Using Hydraulic Model Parameters:

Several studies have explored the application of optimization techniques in leak identification and localization. [2,11,14] incorporated optimization methods upon hydraulic model parameters with minimization between the squared loss of observed and simulated data. By adjusting the model parameters, potential leak locations were detected, enhancing the performance of leak identification in hydraulic systems.

2.2   Data-Driven Approaches with Sequential or Temporal Data:

Data-driven approaches have gained more attention in leak identification systems research. [9] used the K-nearest neighbors (KNN) algorithm, while [10,11,13] implemented convolutional neural networks (CNNs). [8,16] adopted the Bayes classifier, [7,17] applied Long Short-term Memory (LSTM) networks. These data-driven approaches alleviate machine learning techniques to analyze sequential pressure or flow data, enabling the detection of leaks based on arbitrary patterns and deviation analytics.

2.3   The Integration of Bayesian Belief Propagation:

The belief propagation network utilizing Bayesian logic had been widely utilized in various areas of research. [20–25,27] have successfully experimented with this approach. By incorporating expert rules and apriori information, the belief propagation network addresses challenges associated with uncertainties in leak identification and localization, enabling more accurate and efficient detection.
While significant progress has been made in leak identification and localization, challenges still persist. [3,8,12,14] highlight the limited parametric constraints and higher bandwidth utilization as factors hindering the saturation of leak identification systems. So, the integration of supervised ML techniques, such as decision trees, logistic regression, forests, and ANNs must be taken for experimentation to address the challenge (Table 1).

Table 1. Leak factors for Prediction task

References	Impact factors
[29]	pipe fitting, pipe material, ground movement, pipe corrosion, geological condition, traffic loading, water pressure, excavation, pipe age, water temperature, construction quality
[2,12,17,20,21]	construction damage, water hammer, improper pipe, pipe corrosion,, ground collapse, overload and vibration.
[7,12,21,28]	ground conditions, ground movement, excavation, pipe age, temperature, high pressure, pipe defects, quality of workmanship.
[19]	pipe cracked, pipe dislocation, pipe corrosion, pipe length, number of connected pipes
[18]	load vibration , pipe corrosion, ground movement, pipe material, pipe age, construction activities, pipe depth

The analysis of the literature reveals that deep learning architectures were frequent-ly designed to effectively capture intricate data patterns on a large scale, surpassing traditional machine learning models [7] that rely on manually crafted patterns. Artificial neural network (ANN) models have particularly demonstrated superior performance in automatically extracting these patterns, leading to their implementation for estimating combinatorial feature-based multi-thresholding and enhancing prediction accuracy [4]. However, it should be noted that ANNs can also unintentionally model unfavorable data patterns due to their high feature invariance, resulting in over-fitting, prolonged training epochs, and reduced convergence. To tackle this issue, the proposed approach introduces a hybrid model that incorporates Bayesian learning from machine learning. This integration aims to mitigate the impact of temporal streaming data on pattern invariance

3   Proposed Work

The implied approach adopts a control scheme for model-based leak detection and isolation,where feature scoring is generated as the difference between local and global pressure unit variations. The values are calculated by the model in the WDN that takes no-leak conditions into account. The pressure values are assessed when a leak in the WDN is found. It identifies the WDN sub-network that is most likely to contain the leak (Figs. 3 and 4) (Fig. 2).

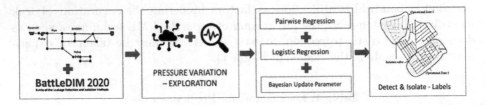

Fig. 2. Architecture of the Proposed Work

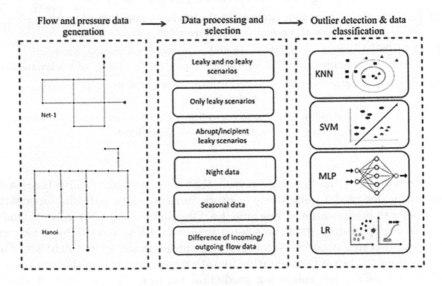

Fig. 3. ML based Leak Identification [12]

A few supervised machine learning techniques (random forest, decision tree, and logistic regression) as well as an artificial neural network (ANN) have been used to conduct the analyses. Below is a brief explanation of the research's methodology. The binary classification method is the logistic regression model [13]. The method taken for this work is driven by the functions:

$$h\_(\theta)(z) = g(\theta \wedge (T) z) \tag{1}$$

$$g(y) = 1/(1 - e \wedge (-y)) \tag{2}$$

where z is the input data and *theta* is the parameter left over after the cost function has been minimized. To choose the model, the decision tree method is combined with a number of decision constraints [10, 12]. This technique divides a sample into divisions that resemble branches and build a hierarchy of nodes. The model builds a flowchart (tree), where each internal node (represented by a question) tests some features based on thresholding and descends through the branches (the outcome of the splitting) with a "gini" coefficient, entropy, or information gain, which is defined as follows:

$$G = \sum_{j=1}^{k} p(i) z (1 - p(j)) \tag{3}$$

The number of classes determines the parameter c, and the probability of selecting a data point belonging to class i is p(i). By combining randomized decision trees, Random Forest methods are used for both classification and regression [13, 16]. A target variable receives support from each decision tree. The combination that receives the most votes is selected by the random forest algorithm. The workings of the human brain are the basis for the ANN [25]. Through a series of internal layers (hidden layers), it transforms and fits the input data (input layer) to the output class layer (output layer). Weights that are modified by optimizing the prediction on training data set serve as the foundation for the fit. For activation, the sigmoid function [4] is utilized.

$$\sigma(z) = \frac{1}{1 + e^{-z}} \tag{4}$$

Where,
z is the input to the sigmoid function; e is the Euler's number (e = 2.781...)

Bayesian Update Parameter Computation
Our aim to improve the accuracy of the hybrid leak localization and detection model computed by below equation. where only the present values of pressure values y^k is considered, As suggested by Xu et al. (2019), archival classification of pressure values can be performed by recursively applying the Bayesian Theorem. The Bayes Theorem

$$P(l_p \mid y^q) = \frac{P(y^q \mid l_p) P(l_p)}{P(y^q)} \quad p = 1 \ldots, n_n \tag{5}$$

where $P(l_p \mid x^q)$ is the posterior likelihood , $P(l_p)$ is the prior likelihood and $P(x^q)$ is a normalization y given by the total probability law.

$$P(y^q) = \int_{p=0}^{n_n} P(y^q \mid l_p) P(l_p) \qquad (6)$$

allows the introduction of previous classification results in Eq. 6 by means the prior likelihood term $P(l_p)$ that can be considered as the posterior likelihood of previous time instant $k - 1$. i.e. consider $P(l_p) = P(l_p \mid x^{q-1})$ in Eq. 5. If Eq. 5 is computed recursively with this determination in a time of H samples, we obtain.

$$P(l_p \mid x^{q-H+n}) = \frac{P(y^{q-H+n} \mid l_p) \, P(l_p \mid y^{q-H+n-1})}{P(y^{q-H+n})} \, , \, p = 1 \ldots, \, n_n, n = 1, \ldots H. \qquad (7)$$

Then, the probable leak localization $\hat{p}$ can be estimated as

$$\hat{p} = \max_{p \,\in\{1,\ldots n_n\}} P(l_p \mid y^q) \qquad (8)$$

where $P(l_p \mid y^q)$ is determined using Eq. 7 in a recursive manner while taking similar probabilities into account for the posterior probabilities at time H=0. When using Eq. 8 instead of Eq. 4, it is possible to filter out some potential timely classification errors caused by uncertainties because the information from values $yk - H + n,\ldots, yk$ is taken into account at time k.

4 Result and Discussion

Evaluation of the proposed work is processed with enhanced BattLeDIM 2020 (Battle of the Leakage Detection and Isolation Methods) dataset. It is an augmented accumulation of different water flow regions (FR1, FR2, FR3 ... FRn) and corresponding pressure values at observation points (PR1, PR2, PR3, PR4). Exploratory analysis is made on this dataset to realize the statistical significance of the dataset as shown in Table 2.

Table 2. Statistical exploratory study of the normalised sample data

Flow region	Min rate (m/s)	Max rate (m/s)	Avg rate (m/s)	SD
FR1	0.14	0.69	0.4	0.17
FR2	0.22	0.61	0.34	0.11
FR3	0.25	0.72	0.47	0.14
FR4	0.59	0.66	0.63	0.09
FR5	0.59	0.71	0.66	0.18

This dataset is spit into train, test parts to feed into the ML pipeline. The experimental results of the proposed work are evaluated using the commonly adopted performance measures, precision, recall, F1-score, and accuracy. The performance is analysed against existing models for combined localization and detection and results are tabulated in Tables 3 and 4.

A series of experiments are conducted with hybrid machine learning and deep learning architectures. Commonly used machine learning models such as KNN, SVM, Logistic Regression etc., are hybridized with ANN and MLP to analyze their potential to estimate temporal data patterns. These traditional models heavily rely on hand-coded patterns, whereas deep learning models, particularly artificial neural networks (ANNs), possess the ability to automatically extract and model intricate relationships from the data. The comparison of deep learning architectures with KNN, SVM, Logistic Regression (LR), and MLP reveals that ANNs, with their flexibility and ability to handle large-scale data, excel in fitting complex patterns more effectively. However, despite the potential benefits of hybridization approaches that combine the strengths of ANNs with other techniques such as KNN, SVM, Logistic Regression (LR), and MLP from machine learning, it is important to acknowledge the better performance of hybrid Bayesian learning with ANN. The hybrid models with KNN, SVM and LR fail to surpass the predictive performance of ANN models in capturing intricate patterns and handling high feature invariance, which can lead to over-fitting and reduced convergence. The integration of Bayesian learning from machine learning into hybrid models can only address the limitations of traditional models like KNN, SVM, Logistic Regression, and MLP in effectively capturing complex data patterns, thereby showing the overall improvement in prediction task. The performance comparison in Tables 3 and 4 using metrics FPR, TPR and Accuracy supports the achievement of performance improvement in the proposed work.

Table 3. Performance Comparison of Base models

Base Models	False Positive Rate	True Positive Rate	Accuracy
Decision Tree	0.90	0.05	0.85
SVM	0.88	0.20	0.87
Logistic Regression	0.78	0.13	0.89
ANN	0.85	0.12	0.93

Table 4. Performance Comparison of Hybrid models

Hybrid Models	False Positive Rate	True Positive Rate	Accuracy
Logistic Regression+BL	0.60	0.18	0.66
SVM+BL	0.92	0.04	0.90
Decision Tree+BL	0.85	0.12	0.92
ANN+BL	0.77	0.14	0.96

Based on the results, performance comparison charts are visually represented for leak detection and leak isolation in Figs. 4 and 5 respectively.

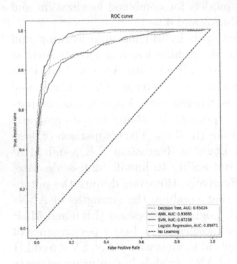 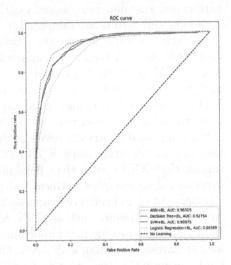

Fig. 4. ROC Curve without BL   **Fig. 5.** ROC Curve with BL

Bayesian learning improves the performance of ML models like ANNs, logistic regression, and SVM by estimating parameters and incorporating prior information. Conversely, the accuracy of the hybrid logistic regression with BL is adversely affected by the temporal apriori, which increases the complexity of the model (leading to overfitting). It handles uncertainty effectively and enables ANNs to make robust predictions by considering prior knowledge and updating beliefs based on observed data. Decision trees benefit from Bayesian learning by adapting to different data distributions and capturing relationships in sequential attributes across temporal spans. Bayesian techniques enhance logistic regression by incorporating prior beliefs and improving parameter estimation, particularly for attributes with temporal dependencies. Bayesian learning also helps SVM models by estimating hyper parameters and adapting to temporal characteristics, resulting in more accurate predictions. Among them, the cumulative thresholding nature of ANNs combined with temporal Bayesian learning (BL) update parameter has brought significant performance improvement with an accuracy of 96% which is 3% higher than its baseline models.

5   Conclusion

Our work proposes a novel technique to predict water leakage challenge in WDNs to support timely detection and isolation of water leaks. The literature study on leakage detection, leak assessment and control show that the variation on the demand and supply of water varies the pressure in the supply or distribution network. Related works on sensing-based simulation and AI supported approaches are abundant in the literature with only little analysis on the pressure dynamism. So, a novel hybrid model is proposed in this paper to detect and isolate leaks by combining Pairwise regression and logistic regression using a Bayesian update parameter to improve the performance of the overall prediction results. Results show that the accuracy of the proposed work increases around 2–3% over existing baselines upon the combined prediction task.

In future, challenges in sensor virtualization, data pre-processing and analytics can be addressed for leak localization, detection and isolation tasks in WDNs.

References

1. Wan, X., Farmani, R., Keedwell, E.: Gradual leak detection in water distribution networks based on multistep forecasting strategy. J. Water Resour. Plann. Manage. **149**(8), 04023035 (2023). https://doi.org/10.1061/(ASCE)WR.1943-5452.0001542

2. Mazzoni, F., Alvisi, S., Franchini, M.: Leakage detection and localization in a water distribution network through comparison of observed and simulated pressure data. J. Water Resour. Plan. Manag. **148**(1), 04021096 (2022)

3. Vrachimis, S.G., et al.: Battle of the leakage detection and isolation methods. J. Water Resour. Plan. Manage. **148**, 12 (2022)

4. Wan, X., Kuhanestani, P.K., Farmani, R., Keedwell, E.: Literature review of data analytics for leak detection in water distribution networks: a focus on pressure and flow smart sensors. J. Water Resour. Plann. Manage. **148**(10), 03122002 (2022). https://doi.org/10.1061/(ASCE)WR.1943-5452.0001597

5. Romero-Ben, L., et al.: Leak localization in water distribution networks using data-driven and model-based approaches. J. Water Resour. Plann. Manage. **148**(5), 04022016 (2022). https://doi.org/10.1061/(ASCE)WR.1943-5452.0001542

6. Ma, X., et al.: A real-time method to detect the leakage location in urban water distribution networks. J. Water Resour. Plann. Manage. **148**(12), 04022069 (2022). https://doi.org/10.1061/(ASCE)WR.1943-5452.0001628

7. Li, Z., Wang, J., Yan, H., Li, S., Tao, T., Xin, K.: Fast detection and localization of multiple leaks in water distribution network jointly driven by simulation and machine learning. J. Water Resour. Plann. Manage. **148**(9), 05022005 (2022). https://doi.org/10.1061/(ASCE)WR.1943-5452.0001574

8. Ravichandran, T., Gavahi, K., Ponnambalam, K., Burtea, V., Mousavi, S.J.: Ensemble-based machine learning approach for improved leak detection in water mains. J. Hydroinform. **23**(2), 307–323 (2021)

9. Levinas, D., Perelman, G., Ostfeld, A.: Water leak localization using high-resolution pressure sensors. Water **13**(5), 591 (2021)

10. Guo, G., et al.: Leakage detection in water distribution systems based on time-frequency convolutional neural network. J. Water Resour. Plan. Manag. **147**(2), 04020101 (2021)

11. Steelbauer, D.B., Deuerlein, J., Gilbert, D., Piller, O., Abraham, E.: A dual model for leak detection and localization. Zenodo (2020). https://doi.org/10.5281/zenodo.3923907. Accessed 15 Mar 2021

12. Kammoun, M., Kammouna, A., Abid, M.: Experiments based comparative evaluations of machine learning techniques for leak detection in water distribution systems. Water Supply. **22**(1), 628–642 (2021). https://doi.org/10.2166/ws.2021.248

13. Fang, Q., Zhang, J., Xie, C., Yang, Y.: Detection of multiple leakage points in water distribution networks based on convolutional neural networks. Water Supply **19**(8), 2231–2239 (2019)

14. Sophocleous, S., Savic, D., Kapelan, Z.: Leak localization in a real water distribution network based on search-space reduction. J. Water Resour. Plan. Manage. **145**(7), 04019024 (2019)

15. Arregui, F.J., Cobacho, R., Soriano, J., Jimenez-Redal, R.: Calculation proposal for the economic level of apparent losses (ELAL) in a water supply system. Water **10**(12), 1809 (2018)

16. Soldevila, A., Fernandez-Canti, R.M., Blesa, J., Tornil-Sin, S., Puig, V.: Leak localization in water distribution networks using Bayesian classifiers. J. Process Control **55**, 1–9 (2017)

17. Leu, S.S., Bui, Q.N.: Leak prediction model for water distribution networks created using a Bayesian network learning approach. Water Resour. Manage **30**, 2719–2733 (2016). https://doi.org/10.1007/s11269-016-1316-8

18. Kuo, T.Y.: Taipei Water Leak Statistics. Taipei Water Company, Taipei (2014)

19. TWC: Taiwan water corporation report (2011)

20. Brady, J., Gray, N.F.: Group water schemes in Ireland - their role within the Irish water sector. Eur. Water **29**(39–58), 2010 (2010)

21. Luu, V.T., Kim, S.-Y., Tuan, N.V., Ogunlana, S.O.: Quantifying schedule risk in construction projects using Bayesian belief networks. Int. J. Proj. Manage. **27**, 39–50 (2009)

22. Martín, J.E., Rivas, T., Matías, J.M., Taboada, J., Argüelles, A.: A Bayesian network analysis of workplace accidents caused by falls from a height. Saf. Sci. **47**, 206–214 (2009)

23. Ching, J., Leu, S.-S.: Bayesian updating of reliability of civil infrastructure facilities based on condition-state data and fault-tree model. Reliab. Eng. Syst. Saf. **94**, 1962–1974 (2009)

24. Beuken, R., LavooÚ, C., Bosch, A., Schaap, P.: Low leakage in the Netherlands confirmed. Water Distrib. Syst. Anal. Symp. **2006**, 1 (2008)

25. Liu, T.-F., Sung, W.-K., Mittal, A.: Model gene network by semi-fixed Bayesian network. Expert Syst. Appl. **30**, 42–49 (2006)

26. Liemberger, R., Marin, P.: The Challenge of Reducing Non- Revenue Water in Developing Countries-How the Private Sector Can Help: A Look at Performance-Based Service Contracting. World Bank, Washington, DC (2006)

27. Antal, P., Fannes, G., Timmerman, D., Moreau, Y., De Moor, B.: Using literature and data to learn Bayesian networks as clinical models of ovarian tumors. Artif. Intell. Med. **30**, 257–281 (2004)

28. Vítkovsky, J.P., Simpson, A.R., Lambert, M.F.: Leak detection and calibration using transients and genetic algorithms. J. Water Resour. Plan. Manage. **126**(4), 262–265 (2000)

29. ISWA: Statistics on impact factors of water leakage. International Water Service Association, IWSA, London (1991)

Real-Time Birds Shadow Detection for Autonomous UAVs

Kassem Anis Bouali[✉] and András Hajdu

Faculty of Informatics, University of Debrecen, Kassai út 26, Debrecen, Hungary
kassemanis@outlook.com, hajdu.andras@inf.unideb.hu

Abstract. Autonomous unmanned aerial vehicles (UAVs) are commonly used for wildlife exploration and animal monitoring. Therefore, bird attacks pose a significant challenge to UAVs. As we know, Traditional Bird Detection methods used for prevention against attacks may fail when the attacks occur from unobservable angles. However, the UAV can gain an early indication of an impending attack if it detects the location of the bird's shadow and takes proactive measures to minimize the risk of damage. To address this, we present the ShadowBirdCUB dataset, derived from the CUB-200-2011 Dataset, which is used to train cutting-edge Deep Learning Algorithms for shadow detection. Experimental results using various Deep learning Object Detection (DLOD) models and performance metrics demonstrate promising effectiveness. Although this approach is limited to detecting attacks from certain angles, it is a valuable addition to existing bird detection methods.

Keywords: UAVs · Bird Detection · Deep Learning Algorithms · DLOD · ShadowBirdCUB · CUB-200-2011 Dataset

1 Introduction

Autonomous unmanned aerial vehicles (UAVs) have become indispensable tools for wildlife exploration and animal monitoring, enabling researchers to study and protect diverse ecosystems and their inhabitants. However, UAVs face a formidable challenge in the form of bird attacks, which can disrupt missions and cause substantial damage to the vehicles. Conventional methods of detecting birds may prove inadequate when attacks occur from angles that are difficult for UAVs to observe. In response to this challenge, we propose a novel approach that leverages the detection of bird shadows to prevent such attacks and safeguard UAVs. By identifying the presence of a bird's shadow, the UAV can receive early warning signs of an imminent attack, enabling proactive measures to minimize the risk of damage. While shadow detection may be limited to certain attack angles, it offers a valuable supplement to existing bird detection methods, enhancing the UAV's ability to mitigate the impact of avian threats.

In this paper, we present the ShadowBirdCUB dataset, specifically curated for training object detectors to recognize and detect bird shadows. Derived from the CUB-200-2011 dataset, the ShadowBirdCUB dataset serves as a benchmark to evaluate the effectiveness of shadow-based detection techniques. Our objective is to assess the performance of various state-of-the-art object detectors using the ShadowBirdCUB dataset. Through comprehensive experimentation, we aim to demonstrate the potential of shadow-based detection techniques in effectively mitigating bird attacks on UAVs. By providing UAVs with the capability to detect bird shadows, we enable them to anticipate and respond to potential attacks, thus enhancing their resilience and ensuring the continuity of their missions in wildlife exploration and animal monitoring.

In the forthcoming parts of this paper, we outline the methodology employed for shadow detection, which encompasses multiple stages. We utilized the SGRNet algorithm to generate synthetic shadows corresponding to the segmented birds from the CUB dataset and discuss the feature extraction methods utilized to extract relevant features from the generated images. Finally, we train object detectors on the processed ShadowBirdCUB dataset and present the experimental results. Overall, our research aims to contribute to the field by introducing an innovative approach to address the challenge of bird attacks on UAVs and demonstrate the effectiveness of shadow-based detection techniques in enhancing their protection and operational integrity. However, it is important to acknowledge the limitations of our method. Firstly, the method relies on the availability of sufficient lighting conditions and specific angles for accurate shadow detection. This means that in certain scenarios, such as low light or when the sun is directly overhead, the effectiveness of the method may be compromised. Additionally, the method is limited to UAVs with single frontal cameras, which are the most common ones, as professional drones with multi-cameras can directly detect the bird itself but are also expensive. Moreover, the shadow-based approach alone may not provide complete coverage for all potential attack angles. Despite these limitations, the shadow-based detection method remains a valuable tool in the UAV's arsenal for bird attack prevention. By supplementing existing bird detection methods with shadow detection, the UAV gains an additional layer of early warning and can take proactive measures to safeguard itself against attacks. This integrated approach enhances the overall protection and operational integrity of UAVs in wildlife exploration and animal monitoring missions. The subsequent sections of the paper are structured as follows. Section 2 is dedicated to related works, where we examine pertinent studies in the literature. The proposed approach is described in Sect. 3 with presenting the methodology utilized to produce our novel dataset and also a comprehensive account of our experimental design. The corresponding results are enclosed in Sect. 4. Lastly, Sect. 5 outlines the conclusions drawn regarding the proposed method and its performance.

2 Related Work

A significant amount of research has focused on detecting shadows in individual images. One approach employed a fusion of adaptive thresholding techniques, which generated a binary shadow mask to identify shadow pixels. However, this method relied on classical image processing algorithms that struggled with complex images containing a variety of dark pixels [16]. On the other hand, Murali S and Govindan VK leveraged the LAB color space, particularly the L channel, to control luminance and accurately detect shadowed pixels, offering unique benefits for shadow detection [11]. Researchers then have turned to Deep Neural Networks (DNNs), including CNN and GAN-based networks, due to their ability to learn and tackle shadow identification challenges [3]. However, these DNNs require a large amount of data, making it challenging to collect sufficient shadow data. As a result, the use of synthetic training datasets has become a crucial method for data gathering. Wrenninge M and Unger J demonstrated the effectiveness of synthetic data by exploring various pre-trained models on their dataset, providing detailed insights into performance variations [15]. In the domain of shadow detection, researchers have synthesized shadows using different image combinations and parameters, introducing large-scale synthetic shadow datasets [6]. Additionally, other researchers isolated shadows from images, incorporated them into different backgrounds using random parameters, and combined them with real data to create synthetic training datasets [10]. Although these methods have made valuable contributions, certain limitations need to be addressed. By employing more advanced shadow generators, improved results can be achieved. Furthermore, the existing approach of shadow extraction and subsequent generation leads to a dataset containing shadows of the same shape, potentially biasing the trained models by not adequately considering variations in shadow shapes. Moreover, the reliance on real data in the shadow generation process poses challenges in terms of data acquisition, which can impede the scalability and accessibility of the method." In our proposed solution, we specifically concentrate on the utilization of pure synthetic data for training advanced bird shadow detectors since gathering real bird shadow data is inherently challenging, which motivated us to leverage such solution. By doing so, we aim to overcome limitations associated with real data usage, and also the variations in the shadow shape and contrast. Furthermore, we incorporated a novel generative neural network SGRNet [5], to enhance the effectiveness of the generated shadows.

3   Proposed Approach

Our objective is to create an exceptional dataset of state-of-the-art bird shadows. To accomplish this, we primarily relied on the following steps: Acquiring a sample dataset, generating shadows, cleaning the data, and conducting post-processing.

3.1   Acquiring a Sample Dataset

We utilized for our task; the Caltech-UCSD Birds-200 (CUB-200-2011) dataset [13], which is a collection of bird images that serves as our sample dataset. It provides a diverse range of bird species and contains a significant number of annotated images. This dataset has also consistently demonstrated effectiveness across numerous tasks [1, 2, 8, 12], which is precisely why we opted to utilize it. Our method is particularly sensitive to the chosen dataset, making this choice more crucial. Additionally, we utilized the binary segmentation derived from the CUB-200-2011 dataset [4] which serves as a valuable resource for the subsequent step of shadow generation.

The chosen sample is limited to on standing and sitting birds mostly, and a relatively small number of flying birds' images, which is logical in a sense, because it is not very easy to take a clean picture of a flying (moving) bird. However, this limitation can affect the diversity of our dataset and reduce the efficiency of our detection model. To resolve this matter, we added some random flying bird images that we collected from google images, which consisted of 5% of the size of the CUB-200-2011 dataset.

3.2 Shadow Generation

We applied a state-of-the-art approach for generating synthetic shadows in real-world scenes which is the SGRNet proposed by Hong, Yan, Niu, Li, and Zhang, Jianfu [5] to our sample dataset. By leveraging SGRNet, with the foreground object mask which is the binary segmentation [4] of our samples, you can modify (resetting the paths of your own data) and run, we were able to generate realistic synthetic shadows on the bird images. Some sample results of the shadow generation process can be observed below (see Fig. 1). However, the approach was not very efficient in terms of generating very realistic looking shadows, most commonly when there is nowhere to reflect the shadow e.g., a picture of a flying bird in the sky (Fig. 2).

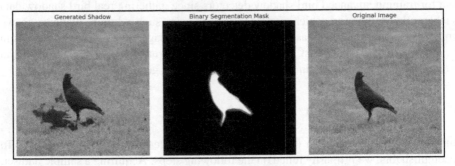

Fig. 1. Shadow generation process using SGRNet.

Fig. 2. Failure analysis of SGRNet: examples of misclassified samples.

3.3 Data Cleaning

After the generation of shadows, it was observed that some of the images were of poor quality and unsuitable for training purposes as we have already mentioned (Fig. 2). To ensure the dataset's integrity, a thorough manual cleaning was conducted to eliminate these inadequate samples.

3.4 Post-processing

We performed post-processing on the generated shadow images to get the shadow annotations. we basically utilized many approaches depending on the type of the data.

At first, we applied an image processing approach: the bird segmentation is used remove the main bird from each image, then the images were converted to grayscale, enabling us to apply a threshold to detect dark pixels associated with shadow regions. After that to extract the shadow contours, we employed the "Find Contours" algorithm and sorted them based on their area in descending order. By prioritizing smaller contours, we focused on capturing the finer details of shadows. It is worth noting that the largest contour, corresponding to the background, was excluded from the grouping process to avoid merging it with the smaller shadow contours. To generate a comprehensive and accurate shadow annotation, we merged the bounding boxes of the smaller contours into a single global bounding box. This bounding box encompassed the collective area of the detected shadows, providing a precise representation of their spatial extent in the image (Fig. 3). While the automated post-processing approach proved effective for some images, it had limitations in detecting shadows in many other cases, especially when the picture is full of dark pixels.

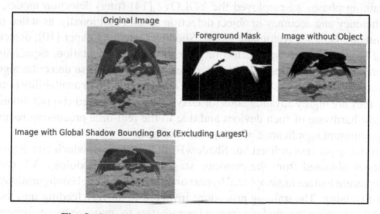

Fig. 3. Image processing method for dataset labeling

This led us to switch to another annotation technique using the SAM (Segment Anything Model) algorithm provided by Meta [9]. We basically extracted the segmentation mask and then linked the edge points to form of a polygon that is used as a label in the case of Polygon annotation (see Fig. 4).

Fig. 4. Labeling approaches demonstration.

However, those annotation technics where not efficient because we cannot distinguish the shadow from the dark areas of the picture in many images. To address this, manual annotation was employed for challenging images, ensuring precise identification and delineation of shadow regions by human annotators. This meticulous process resulted in a high-quality annotated dataset. By combining automated and manual annotations, our methodology achieved a balance between efficiency and precision.

Note: We used a combination of Rectangular and Polygon labeling, but mostly the polygon one especially during manual annotation. Because it gives a very precise annotation, also considering that shadows have very complicated shapes, and it is hard to define it inside a rectangle.

3.5 Detection Model Training

In the training phase, we employed the YOLOv7 [14] (tiny) detection model, known for its efficiency and accuracy in object detection tasks. Additionally, as stated in in the Semi-synthetic Data for Automatic Drone Shadow Detection paper [10], detecting the bounding boxes is easier than performing the shadow segmentation, especially if the images are very complex and full of dark pixels. Moreover, these detection algorithms exhibit notable efficiency in shadow detection [7]. Also given their availability in compact versions, they are highly advantageous for UAVs due to the limited computational power and modest hardware of such devices and due to the real-time processing requirement and the paramount significance of computational time.

Our training process utilized our ShadowBirdCUB dataset, which consists of annotated images obtained from the previous stages of our methodology. We conducted experimentation to determine optimal hyperparameters and model configurations, ensuring robust training. The training procedure involved iteratively feeding into the detection model, adjusting the model's internal parameters to optimize its performance. We demonstrate the performance of the model after 100 epochs of training, where it exhibited exceptional performance and learning convergence, achieving a mean Average Precision (mAP@.5) of 0.43 during a short training time. These results indicate the model's ability to accurately detect and localize bird shadows (Fig. 5) showcasing its potential for effective shadow-based detection in UAV protection and wildlife monitoring scenarios.

95% accuracy   71% accuracy   33% and 56% accuracy

Fig. 5. Performance evaluation of trained YOLO v7 on real captured images.

4 Results and Discussion

We evaluated the performance of the YOLOv7 [14] detection model using different standard metrics (see Table 1). Notably, the model demonstrated promising results, achieving a precision of 70.1% and a recall of 51.3%. The mean average precision (mAP50) at an IoU threshold of 0.50 reached 52%, indicating satisfactory performance across different confidence score thresholds. However, the mAP over a broader range of IoU thresholds (mAP50-95) was 29.8%, suggesting a decline in performance at higher IoU stringency levels.

This observation is reasonable since our evaluation utilized polygon and segmented annotations, whereas object detection models are optimized for precise bounding box annotations. Polygon and segmented annotations define objects at the pixel level, leading to shape variations that can challenge accurate bounding box predictions. Consequently, the mAP metrics are adversely affected since it is impossible to map a rectangular predicted bounding box to a very complicated ground truth-shaped polygon (Fig. 6) resulting in lower overall performance scores.

Still, the model demonstrated excellent performance and high accuracy (see Fig. 5) in real-life scenarios, indicating its ability to detect and localize bird shadows in varying environmental conditions accurately. Furthermore, comparing our shadow detection approach with existing methods, it showcased promising results and outperformed many other models in terms of adaptability, especially for traditional UAVs.

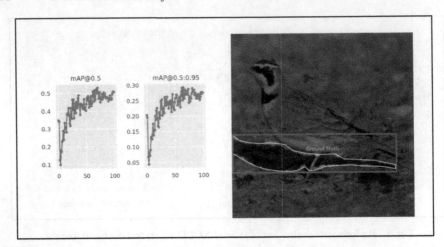

Fig. 6. Impact of Polygon/Segmented labeling on some object detection evaluation metrics.

Table 1. YOLOv7 object detection results on our custom dataset.

Performance metric	Value
mAP50	0.52
mAP50-95	0.30
Box Precision	0.70
Recall	0.51

5 Conclusion

In conclusion, our study introduces a novel approach for UAV protection by leveraging shadow-based detection to mitigate bird attacks. The ShadowBirdCUB dataset, curated explicitly for training object detectors, facilitated our experiments. The results demonstrated the effectiveness of our approach, achieving an impressive precision of 70% and a recall of 51% in detecting and localizing bird shadows. These quantitative results highlight the model's robust performance under varying environmental conditions. While limitations exist, such as occasional false positives, our solution offers valuable enhancements to existing bird detection methods. By anticipating and responding to bird threats with high accuracy, our approach contributes to the resilience and continuity of UAV missions in wildlife exploration and animal monitoring.

Acknowledgment. This work is funded in part by the Scientific Research Fund (OTKA) grant #143540 and by the project TKP2021-NKTA-34, implemented with the support provided by the National Research, Development, and Innovation Fund of Hungary under the TKP2021-NKTA funding scheme.

References

1. Choudhury, S., Laina, I., Rupprecht, C., Vedaldi, A.: Appendix: unsupervised part discovery from contrastive reconstruction In: Ranzato M, Beygelzimer A, Dauphin Y, Liang PS, Vaughan JW (eds.) Advances in Neural Information Processing Systems. Curran Associates, Inc., pp. 28104–28118 (2021)
2. Cole, E., et al.: On label granularity and object localization. In: Avidan, S., Brostow, G., Cissé, M., Farinella, G.M., Hassner, T. (eds.) Computer Vision – ECCV 2022. ECCV 2022. Lecture Notes in Computer Science, vol, 13670, pp. 604–620. Springer, Cham (2022). https://doi.org/10.1007/978-3-031-20080-9_35
3. Dornadula, S.P.K., Brunet, P., Elias, D.S.: AI driven shadow model detection in agropv farms. arXiv (2023). arXiv:2304.07853
4. Farrell, R.: Cub-200-2011 segmentations (2022). https://doi.org/10.22002/D1.20097
5. Hong, Y., Niu, L., Zhang, J., Zhang, L.: Shadow generation for composite image in real-world scenes In: Proceedings of the AAAI Conference on Artificial Intelligence, pp. 914–922 (2022)
6. Inoue, N., Yamasaki, T.: Learning from synthetic shadows for shadow detection and removal. IEEE Trans. Circuits Syst. Video Technol. **31**(11), 4187–4197 (2021). https://doi.org/10.1109/tcsvt.2020.3047977
7. Kaushal, M.: Rapid -YOLO: A novel yolo based architecture for shadow detection. Optik **260**, 169084 (2022). https://doi.org/10.1016/j.ijleo.2022.169084
8. Tibebu, H., Malik, A., De Silva, V.: Text to image synthesis using stacked conditional variational autoencoders and conditional generative adversarial networks. In: Arai, K. (eds.) Intelligent Computing. SAI 2022. Lecture Notes in Networks and Systems, vol 506, pp. 560–580. Springer, Cham (2022). https://doi.org/10.1007/978-3-031-10461-9_38
9. Kirillov, A., et al.: Segment anything. arXiv (2023). arXiv:2304.02643
10. Mokhtari, M.E.A., et al.: Semi-synthetic data for automatic drone shadow detection (2022). https://doi.org/10.14428/esann/2022.ES2022-82
11. Murali, S., Govindan, V.K.: Shadow detection and removal from a single image using lab color space. Cybern. Inf. Technol. **13**(1), 95–103 (2013). https://doi.org/10.2478/cait-2013-0009
12. Verelst, T., Rubenstein, P.K., Eichner, M., Tuytelaars, T., Berman, M.: Spatial consistency loss for training multi-label classifiers from single-label annotations. In: 2023 IEEE/CVF Winter Conference on Applications of Computer Vision (WACV), pp. 3868–3878 (2023). https://doi.org/10.1109/WACV56688.2023.00387
13. Wah, C., Branson, S., Welinder, P., Perona, P., Belongie, S.: Technical report CNS-TR-2011-001, California Institute of Technology (2011)
14. Wang, C.Y., Bochkovskiy, A., Liao, H.Y.M.: YOLOv7: trainable bag-of-freebies sets new state-of-the-art for real-time object detectors In: Proceedings of the IEEE/CVF Conference on Computer Vision and Pattern Recognition, pp. 7464–7475 (2022)
15. Wrenninge, M., Unger, J.: Synscapes: A photorealistic synthetic dataset for street scene parsing. arXiv (2018). arXiv:1810.08705
16. Zou, X., Zhou, R., Helou, M.E., Süsstrunk, S..: Drone shadow tracking. arXiv (2019). arXiv:1905.08214

Knowledge Graph for Fraud Detection: Case of Fraudulent Transactions Detection in Kenyan SACCOs

Ronald Ojino[1]([⊠]) and Raphael Ndolo[2]

[1] Department of Computer Science and Information Technology, The Co-operative University of Kenya, Ushirika Rd, Nairobi, Kenya
ronojinx@gmail.com
[2] Department of Information Technology, United States International University - Africa, Nairobi, Kenya

Abstract. Detecting fraudulent transactions in SACCOs is essential in preventing financial losses and maintaining customer. Many SACCOs incur massive financial losses due to fraudulent activities such as corruption, asset misappropriation and fraudulent financial statements. In response to these challenges, we propose an approach that detects and prevents transaction risk by leveraging knowledge graphs which contain connectivity patterns and relations; combined with rules that are exploited to discover the knowledge between the type of transaction and customer thereby detecting any anomalies. The effectiveness of the approach is evaluated using real-world SACCO transaction data and shows that it can detect potential fraud in real-time or near real-time thereby saving funds that would have been lost.

Keywords: Fraudulent transactions · Knowledge graph · SACCOs

1 Introduction

SACCOs (Savings and Credit Co-operative Societies) play an important part of an economy by affording cheap credit to their members, agro-processing, marketing of agricultural produce, creation of employment opportunities. The SACCO movement in Kenya is the largest in Africa (contributing approximately 20% of the country's domestic savings) and among the top ten globally [1]. The continuation of the SACCO movement is bound to increase the net size of financial inclusion by bringing the previously excluded majority [2] who were left out by financial institutions like commercial banks. However, this movement just like others in the financial services sector, has been bedeviled by various forms of fraud such as corruption, asset misappropriation and fraudulent financial statements carried out by employees, customers, management or other external parties. The occurrence of numerous cases of fraud in Kenyan SACCOs has led to the loss of revenue and assets [3]. This calls for a raft of measures to detect fraud in order to ensure the sustainability of this sector of the economy.

S. Tiwari et al. (Eds.): AI4S 2023, CCIS 1907, pp. 178–186, 2023.
https://doi.org/10.1007/978-3-031-47997-7_14

Fraudulent activities in SACCOs are buoyed by the evolving complex methods of doing business such as mobile transactions which are risky in nature. In order to address the fraud menace, Financial Accounting Information such as disclosure, transparency, dissemination and clarity is required [4]. Such measures are enforced via internal controls like the control environment (e.g. proper staff recruitment, existence of well formulated policies and better remuneration), risk assessment, control activities (e.g. verification before making payments and supervision), evaluation of the internal control system via audits, inspections and peer reviews, and the implementation of an open and effective ICT system. However, [5] note that internal controls among SACCOs is poor and remains a capable culprit in fraud cases.

ICT tools can be applied to effectively detect, predict and prevent fraudulent transactions such as payment fraud, money laundering, identity theft etc. through technologies like machine learning, deep learning and knowledge graphs. In this paper, we focus on the use of knowledge graphs in fraudulent transaction detection in SACCOs. Knowledge graphs being representations of a set of interconnected facts and concepts within a domain provide ways to organize and understand large amounts of data by linking concepts together in a way that reflects the relationship among them and can be exploited in order to detect and prevent fraud. They can mine valuable hidden data from a large scale associated data as a new form of knowledge representation [6]; they achieve this by handling and structuring all sorts of data in turn uncovering hidden relations between the data items [7]. Furthermore, they provide better performance for certain types of queries which involve relationships in comparison to traditional databases that rely on tables and join operations. This study presents a knowledge graph based model of financial fraud transactions that allows the representation of each banking transaction, customers and financial entities as nodes and having properties that provide additional information about the node, as well as vertices that represent the connections between nodes.

The rest of the paper is organized as follows: Sect. 2 discusses related work of fraud detection using knowledge graphs and ontologies. Section 3 illustrates our pro-posed methodology of using knowledge graphs in fraudulent transaction detection in SACCOs. We present the experimental results of using the proposed approach in Sect. 4 which is succeeded by concluding remarks and a road-map for future work in Sect. 5.

2   Related Work

SACCOs have numerous transaction types that occur between the SACCO and its clients or with other financial institutions. Nowadays most SACCOs provide online transactions due to the convenience they offer customers, and ability to reach wider markets. Transaction fraud is not limited to a geographical boundary such as a county but instead can occur from any location based on the global nature of online transactions. These transactions provide large amounts of data in a given financial period but are challenging to transform into useful knowledge that can help combat fraud. With the increasing amount of online transactions which add efficiency and transparency, there is need for sophisticated techniques to identify fraud [8] given that such transactions can harbor loads of discrepancies attributed to fraud. For instance popular large language models can be used for the construction of knowledge graphs [9] such as the ones used in fraud detection. Fraud detection systems are categorized into two, in e-government systems i.e. those that detect fraudulent activities the minute they take place and those that identify fraud by discovering suspicious behavioral patterns within batches of data [10]. Manual based methods of mining information from large amounts of data e.g. in fraud detection suffer from struggling to handle increased data volumes and heterogenity [11].

Among the first efforts in applying knowledge graphs for fraud detection was by [10] who developed a fraud ontology for fraud detection in e-governments. Ontologies are the basis of knowledge graphs giving a representation structure of items and reasoning in a domain that make reasoning explicit [12]. Knowledge graphs are designed to excel at managing highly connected data and handling queries that involve traversing relationships. They enable efficient navigation of complex networks of data, making them well-suited for fraud detection. Ontologies being the basis of knowledge graphs impact the type and quality of reasoning that can be made by a knowledge management system through making explicit implicit knowledge. The knowledge base for fraud helps in understanding the domain and bridges the gap among different stakeholders by offering a common understanding. The generic fraud ontology can be used as the basis in building case specific fraud ontologies and other fraud applications as it provides useful insights in performing knowledge modeling in specific scenarios.

[8] developed a Knowledge Base ontology for fraud detection using topic modeling from a range of news articles. They analyzed key phrases using TF-IDF scores and then applied Latent Dirichlet Allocation (LDA) to extract relevant key phrases which were used for ontology creation. Whenever a new transaction occurs, it is stored in an ontology graph and then it is compared with the knowledge base to determine the type of fraud in the transaction. A key concern however is that the fraud ontologies put forward by the studies are not publicly available for reuse.

Some authors have improved fraud detection by integrating ontologies and machine learning algorithms. [13] present a knowledge based system for financial fraud detection based on an ontology, SWRL and a decision tree algorithm in order to provide early warning to regulators and support investor decision making processes. The system presents an ontology of financial statements, and utilizes the decision tree algorithm to find financial statement fraud patterns and transforms those patterns to SWRL rules. The SWRL rules and ontology normalize the knowledge of fraud activities and create a knowledge base for financial state. [6] use knowledge graphs to detect and prevent Related Party Transactions (RPTs) which are fraudulent. They posit that if financial fraud detection only considers individual companies' data and ignores information about RPTs, fraudulent transactions will slip through the net. The knowledge graph was able to detect static relationships and dynamic relationships such as transactions between related party and affiliated listed companies, related party and non-affiliated listed companies, and the related parties themselves with results showing that the method improves financial fraud detection. Furthermore, machine learning methods were applied to verify and improve the performance of the method. [14] design a KG framework that discovers knowledge from the relationship between managers and the associated institutions to enhance fraud detection. They use machine learning algorithms to evaluate the classification performance of framework and found support vector machine (SVM) to be the best.

Based on the above, it is evident that knowledge graphs continue to play a key role in aiding the detection and prevention of fraudulent financial transactions across a range of sectors.

3   Proposed Approach

The data used in this research was obtained from a leading SACCO's core transaction system. The data is stored in multiple tables but of interest is the main transaction table and reference data table. The transaction table contains all transactions carried out by the different branches of the SACCO (see Table 1) whereas the reference data table is a table that contains all the transactions that have been investigated and con-firmed to be fraudulent by the SACCO's financial analysts (see Table 2).

To detect, predict and prevent fraud, a graph based approach was utilized on the SACCO's dataset. This approach was preferred given the underlying relationships between entities in the provided data that can be inferred to detect and display anomalies in the dataset. Through the approach, the dataset was converted into a corresponding knowledge graph representation. Neo4J database was used to store, manage, and query highly connected SACCO transactional data using a graph-based model. Neo4J was selected for use due to its performance, scalability, and expressive query language and that is the reason we decided to use the database. Data was collected from the SACCO for a period of 3 weeks as from 10th January 2022 to 31st January 2022. Due to large number of transactions performed in a SACCO during a day, the dataset was deemed sufficient for use in this work. The knowledge graph fraud detection pipeline is depicted below (see Fig. 1).

The database provides a native graph storage and processing engine, enabling it to handle large amounts of data with complex relationships effectively. Neo4j's query language Cypher was used for traversing and querying the graph patterns and this is particularly useful for fraud detection, as complex queries that traverse the graph to uncover suspicious patterns, relationships, or sequences of events can be expressed. For example, patterns such as multiple accounts linked to a single customer, unusual transaction flows, or connections between seemingly unrelated entities can be visualized.

Table 1. Transaction table

Field	Description
Source_cif_id	This is a unique identification number assigned to a member. A member can have different accounts but all fall under one cif_id which uniquely identifies a member. The cif_id represented belongs to source account
Destination_cif_id	This is the cif_id for the destination account
Source account	Refers to the account number in which money was credited from
Source amount	This is the amount of money that was credited from the source account
Source currency	Is the currency that was used to transact funds from the source account
Destination account	This is the account number into which money was debited
Destination amount	This is the amount of money that was debited to the destination account
Destination currency	This is the currency that was used to transact funds to the destination account
Transaction id	Refers to a unique identification number used to identify a specific transaction
Transaction date	This is the date on which a transaction took place
Transaction status	Refers to the status of a specific transaction. The transactions have the following statuses, new, closed, under investigation. When a transaction is created, it is assigned the status new. When analysts pick it up to be investigated, it has the status under investigation. Once all the investigations have been carried out and concluded, the transaction had the status closed
Transaction channel	This is the channel through which the transaction took place. The SACCO has several channels including USSD, web, mobile application, and branch
Transaction risk score	This is a score generated by the rule engine, to rate a transaction against the business rules. Once a transaction has a score of more than or equal to 100, It is pushed to the transaction table for further investigation
Created on	The date and time that a transaction was recorded into the database
Updated on	The date and time that a transaction was updated in the database

Table 2. Reference data table

Field	Description
Transaction id	Foreign key to the transactions table
Created on	Date and time that a transaction was recorded into the database
Updated on	Date a transaction was updated in the database

4   Results

The developed knowledge graph's performance and scalability make it suitable for real-time or near real-time fraud detection scenarios. It can handle large volumes of transactional data and rapidly process queries to identify potential fraud as it occurs. By analyzing the graph in real-time, suspicious activities can be detected in real time and appropriate actions taken to mitigate against potential losses. Sample rules used to detect fraud in the Sacco as written in Cypher are outlined in Table 3 below:

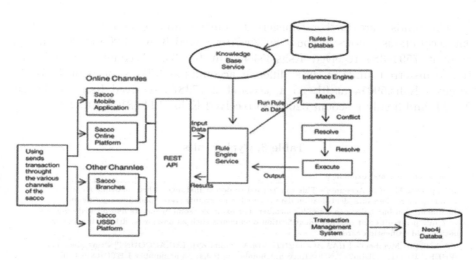

Fig. 1. Proposed fraud detection pipeline in SACCOs

4.1 Sample Fraudulent Funds Movement Detection

From Fig. 2 below, funds transfer movement from one account to another can be clearly seen. Using a Knowledge graph in Neo4J, the movement of funds from one account to another can be traced and the rules detected that the movement of the funds to many accounts within a short duration was suspect. The graph depicts a fraudulent transaction that was carried out on 16th January 2022 at 00:33 h, and by 01:30 h, the cash had been transferred from one account to fourteen other accounts. The initial transaction saw Kshs.712,772.00 moved from an account with the CIF ID 2549059176938. Since this was an unusual withdrawal from this account, the transaction was picked by the organization's rule engine and was reported for further investigation. However, once the funds were transferred to 2549059176938, it was noted that the funds were transferred to seven different accounts in less than ten minutes. from the initial transaction the funds were transferred as follows: account 31851814 received Ksh. 95,900 at 00:40 h, account 45997788 received Kshs.90,000 at 00:42 h, account 77455547 received Kshs.82,623 at 00:43 h, account 59544830 received Kshs.88,993 at 00:45 h, account 30211010 received Kshs.104,638 at 00:47 h, account 54607754 received Kshs.296,753 at 00:50 h and finally account 92234998 received Kshs.46120 at 00:52 h. After receiving Ksh.95,900, account 31851814 transferred Ksh.90,000 to account 45997788 at 00:42 h.

The funds were further transferred from account 45997788 to two different accounts as follows: account 89929679 received Kshs.44,900 at 00:54 h and account 75944385 received Kshs.354,483 at 01:33 h. Account 75944385 further transferred funds to three different accounts as follows: account 41670215 received Kshs.59014 at 01: 35 h, account 23173873 received Kshs.150,487 at 01:37 h and finally ac-count 32068485 received Kshs. 144,681 at 01:39 h.

Table 3. Cypher rules

Rule description and example
a. Duplicate Member Accounts - This rule is used to detect instances where multiple member accounts share the same identification details or where accounts were opened using different documentation but belong to a similar member. For example, some members had opened different accounts using different national identification documents such as passports, driving licenses or national Identification cards MATCH (m1:Member) - [:HAS_ACCOUNT]->(a:Account) <-[:HAS_ACCOUNT] - (m2:Member) WHERE ID(m1) <ID(m2) AND m1.account_number = m2.account_number RETURN m1, m2
b. Unusual member account access. This happens when a member account is accessed from multiple locations within a very short amount of time. MATCH (m:Member)-[r:ACCESS]->(a:Account) WITH m, a, COLLECT(r) AS accesses WHERE size(accesses) >1 AND all(x IN accesses WHERE abs(duration.between(x.timestamp, accesses[0]. timestamp).seconds) <threshold) RETURN m, a, accesses
c. Excessive loan delinquency. This rule helped in identifying individuals who might be aiming at intentionally defaulting on loans or engaging in fraudulent borrowing. MATCH (m:Member)-[r:HAS_loan]->(l:Loan) WHERE l.status="delinquent" WITH m, COUNT(r) AS Delinquency_Count WHERE Delinquency_Count>{threshold} RETURN m, Delinquency_Count
d.Unusual account activity. This rule was used to monitor sudden spikes in account balances resulting from high transaction volumes, or unexpected transfers between member accounts. With this rule in place, cases of potential money laundering, unauthorized account access, or fraudulent fund transfers can be identified. MATCH(a:Account)-[r.TRANSACTION]->(b.Account) WITH a, sum(r.amount) AS totalTransactionAmount WHERE totalTransactionAmount >{threshold} RETURN a, totalTransactionAmount
e. Unusual Saving and withdrawal behaviors. This rule was used to detect cases of money laundering in events where members consistently made large deposits or withdrawals from their accounts, which deviated significantly from their usual patterns. MATCH(m:Member)-[r:TRANSACTION]->(a:Account) WITH m, a, sum(r.amount) AS totalTransactionAmount WHERE totalTransactionAmount >{threshold} RETURN m, a, totalTransactionAmount
f. Identity verification of members was a rule put in place to aid in verifying member identities during member registration, loan application or any transaction associated with the SACCO. The rules included validating identification documents and cross checking all the member information provided to detect and prevent identity theft or cases of fraudulent membership. MATCH(m:Member)WHERE exists(m.identification_document) AND exists (m.first_name) AND exists(m.last_name) AND exists(m.dob) AND exists(m.email) RETURN m

The working of the knowledge graph was demonstrated to 5 financial analysts who were later given a survey to report on its performance. The results of the survey con-firmed that the Knowledge graph dis efficient in flagging fraudulent transactions in a near real-time manner.

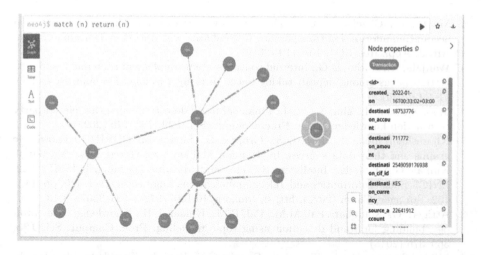

Fig. 2. Knowledge graph of fraudulent funds movement

5 Conclusion and Future Work

Graphs have immense benefits such as explainability, handling of feature inconsistency, exploration and anomaly detection etc. which have increased their popularity. This is one of the first studies that exploits knowledge graphs and applies them in fraud detection within SACCOs. By having real time system checks such as the fraud detection knowledge graph which uncover hidden patterns in data such as tightly knit commmunities like fraud rings, SACCOs may limit the extent to which fraud may take place and thus realize the goal of economic empowerment with reduced forms of corruption. The knowledge graph presented in this paper can detect fraud in real-time or near real-time as it can rapidly process queries. Such a method is much better than the traditional fraud detection patterns as it can learn complex patterns. In future work, the developed knowledge graph can be augmented with advanced analytics and powerful machine learning algorithms without distracting the underlying pipeline to enhance fraud detection capabilities. We also hope to apply such a knowledge graph in detecting fake reviews, fake accounts and disinformation in social media as well as other forms of financial fraud.

References

1. Koskei, I.: Factors influencing the type and occurrence of fraud in deposit taking Sacco's in Kenya. Ph.D. thesis, Strathmore University (2019)
2. Wamukota, M., Ondiek, B., Musiega, M.: Effect of accounting information and communication control on financial performance of Sacco's in Kenya (2022)
3. Kamau, E.N.: An investigation into the causes and characteristics of fraud in Kenyan Sacco and whether Bedford's law can be used to detect fraud in the accounting data. Ph.D. thesis, Strathmore University (2016)

4. Mwita, R., Chachage, B., Mashenene, R.G., Msese, L.: The role of financial accounting information transparency in combating corruption in Tanzanian Sacco. Afr. J. Appl. Res. **5**(1), 108–119 (2019)

5. Wanjala, K., Riitho, D.G.: Internal control systems implementation and fraud mitigation nexus among deposit taking Sacco in Kenya. Financ. Econom. Rev. **2**(1), 11–29 (2020)

6. Mao, X., Sun, H., Zhu, X., Li, J.: Financial fraud detection using the related-party transaction knowledge graph. Proc. Comput. Sci. **199**, 733–740 (2022)

7. Khorashadizadeh, H., Tiwari, S., Groppe, S.: Survey on COVID-19 knowledge graphs and their data sources. In: Nandan Mohanty, S., Garcia Diaz, V., Satish Kumar, G.A.E. (eds.) Intelligent Systems and Machine Learning. ICISML 2022. LNICS, Social Informatics and Telecommunications Engineering, vol. 470, pp. 142–152. Springer, Cham (2022). https://doi.org/10.1007/978-3-031-35078-8_13

8. Attigeri, G., Manohara Pai, M.M., Pai, R.M., Kulkarni, R.: Knowledge base ontology building for fraud detection using topic modeling. Proc. Comput. Sci. **135**, 369–376 (2018)

9. Mihindukulasooriya, N., Tiwari, S., Enguix, C.F., Lata, K.: Text2kgbench: a benchmark for ontology-driven knowledge graph generation from text. arXiv preprint arXiv:2308.02357 (2023)

10. Alexopoulos, P., Kafentzis, K., Benetou, X., Tagaris, T., Georgolios, P.: Towards a generic fraud ontology in e-government. In: International Conference on E- business, vol. 2, pp. 269–276. SCITEPRESS (2007)

11. Zope, B., Mishra, S., Tiwari, S.: Enhancing biochemical extraction with BFS-driven knowledge graph embedding approach. (2023)

12. Ojino, R.O.: Towards an ontology for personalized hotel room recommendation: student research abstract. In: Proceedings of the 35th Annual ACM Symposium on Applied Computing, pp. 2060–2063 (2020)

13. Tang, X.-B., Liu, G.-C., Yang, J., Wei, W.: Knowledge-based financial statement fraud detection system: based on an ontology and a decision tree. KO Knowl. Organiz. **45**(3), 205–219 (2018)

14. Wen, S., Li, J., Zhu, X., Liu, M.: Analysis of financial fraud based on manager knowledge graph. Proc. Comput. Sci. **199**, 773–779 (2022)

Conceptual Framework for Representing Knowledge in the Energy Sector

Sarra Ben Abbes[✉], Lynda Temal, Oumy Seye, and Philippe Calvez

CSAI Lab ENGIE, Paris, France
benabbessarra@gmail.com, fx.research@proton.me

Abstract. This paper aims to contribute to the harmonization and representation of semantic data models used in future energy management systems. In order to support interoperability and maintain data coherence, it is crucial to interpret exchanged information consistently and accurately. Common semantic data models play a vital role in acquiring semantic knowledge and addressing the challenge of semantic interoperability. These models enable effective interoperation between heterogeneous systems in terms of both data and services. This paper focuses on our experience in the Platoon H2020 EU project and its seven pilots. Within this context, we provide background information on data models, present an overview of existing semantic data models, describe the proposed methodology for this task, and provide a detailed account of each step in the methodology. The modeling process was conducted in two phases: the first phase involved scenario descriptions, while the second phase utilized datasets from various pilots.

Keywords: Energy domain · Ontology Modelling · Design Methodology · Design pattern

1 Introduction

The energy sector is currently undergoing its most significant transformation since its inception. It is expected that energy consumption will experience exponential growth over the next decade. This anticipation has prompted actors in the energy sector to shift towards decentralization and decarbonization. Simultaneously, the digitization of the energy sector is rapidly progressing, facilitated by the deployment of sensors and IoT platforms across all components of the energy network. These technologies capture vast amounts of data. Additionally, advancements in computing capabilities have provided the means to extract value from data and convert it into knowledge.

However, to date, different players have been hesitant to share their data, primarily due to the absence of a common legal framework that enables controlled and effective data sharing. Furthermore, owners and operators of energy infrastructures express the need for energy-specific tools that can extract knowledge from sensor-collected data for informed decision-making, enabling them to align their businesses with the evolving dynamics of the energy sector.

S. Tiwari et al. (Eds.): AI4S 2023, CCIS 1907, pp. 187–201, 2023.
https://doi.org/10.1007/978-3-031-47997-7_15

In this context, the PLATOON (Digital PLAtform and analytical TOOls for eNergy) project, with 20 European partners, aims to create a digital platform that facilitates the exchange of data for large-scale multi-party collaboration. Consequently, ensuring semantic interoperability emerges as the most significant challenge to overcome. The partners deal with a wide spectrum of heterogeneous data sources, formats, and interfaces.

This paper primarily focuses on providing feedback on the design of the common semantic data model. We outline the methodology employed to design the semantic data models for the PLATOON H2020 project, based on 19 defined use cases within the seven pilots executed by different partners. The paper is structured as follows: Sect. 2 discusses existing methodologies for building ontologies, Sect. 3 explains the methodology used to design the PLATOON common semantic data model, Sect. 4 illustrates the overarching relationships between the seven pilots, Sect. 5 elaborates on the application of the methodology across the seven pilots, Sect. 6 engages in a discussion regarding the project's objectives and the obtained results, and finally, Sect. 7 concludes the paper.

2   State of the Art

Developing ontologies is a fastidious and time-consuming task for which there exists only some very general principles to guide the ontology designers [6,8] who still must face many modeling choices. Indeed, there is not a single, consensual ontology-design methodology. Several existing methodologies have been proposed such as:

- Cyc methodology [5]: presents to process a large amount of common-sense knowledge which is being built upon a core of over a million hand-entered assertions designed to capture a large portion of what people normally consider consensus knowledge about the world. Three phases are defined: (i) the first phase proposes manually coding the explicit and implicit knowledge appearing in the knowledge sources, (ii) the second phase proposes knowledge codification, and (iii) the third phase delegates most of the work to the tools.
- Hitzler *et al.* [4]: presents a tutorial that is based on use case studies and design pattern application and follows different steps from existing ontology design methodologies.
- KACTUS methodology [7]: this methodology is proposed as a part of the Esprit KACTUS project. The main objective of this project is to investigate the feasibility of knowledge reuse in complex technical systems and the role of ontologies to support it.
- Methodology of Uschold and King [10]: this methodology aims to develop and evaluate ontologies. It includes different stages such as identifying the purpose, building the ontology, ontology coding, integration of existing ontologies, evaluation, etc.
- METHONTOLOGY [2]: this methodology is developed within the Laboratory of Artificial Intelligence at the Polytechnic University of Madrid. It aims to construct ontologies at the knowledge level. This methodology considers

relationships between the life cycle of different ontologies. It is also included detailed recommendations for re-engineering ontologies.

- Methodology of Bravo *et al.* [1]: this methodology promotes the reutilization of ontologies by implementing ontology modules from the beginning of the ontology design, ontologies are seen as reusable modules, not as general design patterns. It is based on 4 main steps: (i) requirements specification, (ii) formal design, (iii) construction, and (iv) evaluation.
- SENSUS methodology [9]: it is based on the use of a huge ontology for building specific ontologies and knowledge bases to be used in applications. SENSUS is an ontology to be used in natural language processing.
- TOVE [3]: this methodology is defined to build an ontology for the TOronto Virtual Enterprise (TOVE) modeling project. The TOVE ontology was constructed to represent a common-sense enterprise model.

In the state-of-the-art, there is no consensus on which methodology is better than the other. Different input parameters can determine the choice of methodology adopted to design ontologies. Each group uses its methodology, and it is odd to find someone who uses a methodology elaborated by a different group. In our work, we are inspired by the methodology described in [1,4] and adapt some steps to fit our need for describing use case requirements. Thus, we mainly use the bottom-up methodology, staring from the scenarios, extract of data and the needs expressed by experts.

3   Methodology for Semantic Data Model Design and Construction

To design the semantic data model for the PLATOON project, we define a methodology that allows us to reach the requirement of each LLUCs (Low-Level Use Case) defined by its different pilots. The Methodology is inspired by several works in the state-of-the-art of ontology design domain [16;18]. However, the difficulties are related to the large number of use cases and the specific need for each use case that furthermore can concerns several transversal domains.

The methodology is divided into four steps detailed below. Each step is applied to each LLUC to cover all the specified needs (see Fig. 1).

3.1   Step 1: Ontology Requirements Specification

Before starting to design ontology, it is important to start to analyze the requirement and the specification expected from the ontology or semantic data model. This step1 is composed of different important tasks to meet the ontology requirement. These tasks are detailed below (see Fig. 1):

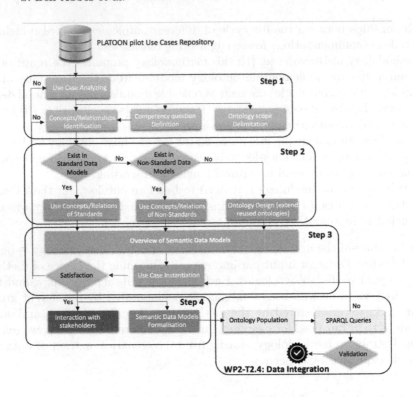

Fig. 1. Semantic Data Model Methodology Design's Steps

(a) **Use case and analyzing**: First, a deep analysis of the requirement (Use Case) is mandatory. In the PLATOON project, each business use case was described in two documents:

   (a) the PM2 template where a High-level description, was given for a specific pilot

   (b) the IEC-62559-based document that describes in detail each Low-Level Use Case (LLUC) for each pilot. The analyzing phase leads to having a big picture and helps to assess the motivations and objectives for each pilot. Details about the domains, actors, agents, services, datasets, objectives, interactions, etc., are given by the IEC-62559.

   However, often use case IEC-62559 analyzing is not enough and should be completed by analyzing an extract of the database related to the use case if provided.

(b) **Ontology scope delimitation**: The delimitation of the scope of ontology is very important to design a modular ontology. Indeed, often use cases are cross domains and because we cannot define everything in the world, an ontology should have a limited scope that can facilitate its definition and sharing. Then, in this task, the scope of the ontology is defined according to the application domain. An example of scope limitation could be Building Energy efficiency and Heating and Cooling Systems.

1. **Competency questions definition**: are questions in natural language that domain experts want the ontology help to answer. The knowledge engineer together with the group of domain experts should produce a list of competency questions. Such competency questions are generated by asking the group of domain experts to enunciate direct questions that they expect the ontology system will be able to answer once it is implemented and in production. The list of competency questions will also be useful to evaluate the final ontology. An example of competency questions can be:
   - What are the sensors used for energy consumption in building x, and what are the types of these sensors?
   - What is the energy consumption of each system contained in a building x?
   - What is the occupancy of a zone and what is the forecast of the occupancy?

(c) **Term elicitation**: to do this task, the knowledge experts analyze the IEC-62559 documents that describe the low-level use cases and identify and extract all terms or notions that are relevant for a particular domain. Furthermore, they also analyze the list of competency questions to extract key terms relevant to be included in the semantic model. Examples of key terms are sensors, building, energy consumption, what (type of sensors), and where (sensors are located).

In summary, the output of this step 1 is a template document with the scope of the ontology, a list of competency questions, and a tab with a list of extracted relevant terms. These terms were analyzed with the domain business experts to validate if the terms extracted are effectively relevant to be designed in the semantic model. Furthermore, the domain business experts can add terms in the tab of the extracted terms if needed to cover notions missed in the IEC-62559 document.

3.2   Step 2: Ontology Analysis

Step 2 aims to create a socle to build the semantic data model. Following the principle of the ontology domain, ontology reusing is recommended before starting any new design. According to this principle, we define several tasks detailed below:

(a) **Identification of concepts and relationships**, from the list of extracted terms during the elicitation task of step 1, we associate for each term or a key notion a concept name or a relation name that will be used in the semantic data model. For example, we associate a concept Sensor with the term sensor and the concept HumiditySensor with the term humidity sensor.

(b) **Reusing Ontology** as previously noticed, reusing ontology should be privileged before creating new concepts. For each identified concept in the previous task, we search in existing ontology if this concept is already defined. When the concepts exist in several ontologies, our strategy is to propose a mapping between these equivalent concepts to ensure interoperability. However, before effectively reusing a concept we analyze the hierarchy of the concept to be sure that the subsumption relation (is-a) is not confused with the part-whole relationship (part-of), composition, or location relationships.

(c) **Extending Ontology** in the second case, the ontological module in the domain exists and covers a part of the use case but the concept doesn't exist in these modules. In this case, we extend the existing ontology with the new concept.

(d) **Ontology Construction** in the third case, the existing ontologies don't cover the use case domain. In this case, we create new ontological modules that cover the use case. However, we extend or add equivalent concepts when it is relevant to increase interoperability.

In summary, the output of this step 2 is: (i) a list of identified existing ontological modules, (ii) a list of reused concepts and relations, and (iii) a list of concepts that need to be designed in a new ontological module to cover the scope of the LLUC.

3.3   Step 3: Overview of Ontological Modules

The third step of the proposed methodology aims at integrating all modules together into a harmonized semantic data model and producing an example for each pilot. This phase consists of the following procedures: diagrams integration, ontology evaluation (scoping of Use Case, consistency, competency), and pilot instantiation with an illustrative example. This step takes as input: (i) a list of identified ontological modules, (ii) a list of concepts and relations coming from the list of ontological modules, and (iii) a newly designed ontological module.

(a) **Diagrams integration** the knowledge engineer puts different modules together in a schema diagram to check all classes, and properties and possibly to improve them. This task is important to join the different schema diagrams of modules with semantic relations (e.g., subsumption, equivalence, etc.). A global overview schema of all used modules should be provided (see Fig. 1).

(b) **Ontology evaluation (scoping of Use Case, consistency, competency)** The knowledge engineer will evaluate the ontological modules to ensure that their definitions correctly implement the use case requirements and competency questions. The goal of ontology evaluation is to prove compliance of the world model with the world modeled formally. Two important aspects are used for evaluation:
   - Competency of the ontology: verify that a representational model is complete concerning a given set of competency questions.
   - Quality requirements: can be measured as the degree of compliance it has to established design criteria (Clarity, Coherence, Modularity).

(c) **Use Case instantiation** with an illustrative example the knowledge engineer instantiates scenarios of the use case with an illustrative diagram

3.4   Step 4: Interaction with Stakeholders and Ontology Formalization

The objective of the fourth step of the methodology is to interact with stakeholders and code all ontology modules by using an ontology editor and a standard language and integrate all modules into an ontology system. This step consists of the following tasks: (*i*) discussion with stakeholders and (*ii*) ontology formalization.

(a) **Discussion with stakeholders** If the stakeholder is unsatisfied, the knowledge engineer lists a set of issues and returns to Step 1.
(b) **Ontology formalization Process** If the stakeholder is satisfied, the knowledge engineer proceeds to the formalization process:
  – provides an owl file for each new ontological module if created in Step 2
  – provides an RDF file for each use case, describing how the overview model is instantiated. This file should be taken as an example to feed the knowledge base

4   Overview of Main Pilots' Topics

The PLATOON H2020 project encompasses seven pilots concerning 19 use cases in total. The project was huge and the use cases were different with some similarities. The description of each pilot was oriented on proposing innovative and smart services in the energy domain that include multiple fields and systems such as wind farms, buildings, smart micro grid and electricity balance, etc.

Analysing pilots use cases, bring us to assess that pilots and use case can have specifics needs, but use cases can evolve in overlapping domains. Indeed, these pilots share several common notions in different domains. Thus, in order to have an harmonized design, determining the set of overlapping concepts that are similar in meaning and are unique to each use cases become essential. This step is important in our process to create harmonized semantic data models that include the information of all the use cases. From the analysis of these pilots, we have roughly identified five topics (see Fig. 2).

– Topic 1 (orange circle) concerns the building, its zones, and the associated other types of building such as retail, logistic centre, data centre, etc. This topic is also related to the properties of a building for instance, energy load, gas, or electric consumption
– Topic 2 (green circle) concerns the HVAC equipment, its subsystems such as heating, cooling and ventilation system. It is also related to the Air Handler Unit
– Topic 3 (blue circle) concerns smart/microgrid, electricity generation, balance, storage, and its properties.
– Topic 4 (grey circle), concerns wind turbine, its components and Photovoltaic plant.
– Topic 5 (white circle), concerns common notions such as sensors and meters, weather, schedule, failures, etc.

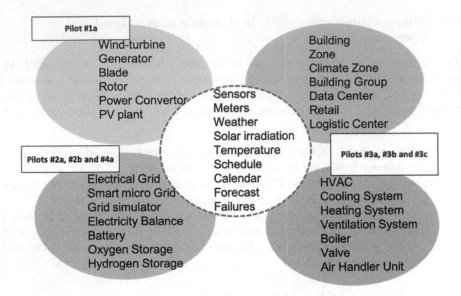

Fig. 2. Overview of main topics

Pilots #3a, #3b and #3c share 2 topics (topic 1 and topic 2) related to Building and HVAC domain. Pilots #2a, #2b and #4a share topic 3 related to the Grid domain. The pilot #1a is in an independent topic (topic 4) related to the energy renewable but could share some notions on power production and electricity generation. All pilots share common concepts (topic 5) which are related to the weather, measurement, forecasting, damages, and failures, etc.

5   Methodology Application

This section aims to describe case studies of the application of the proposed methodology for all the pilots that are presented in PLATOON. We defined a common template that is provided to the different partners to correctly define the scope of the domain ontology and to easily exchange with the stakeholders.

5.1   Application of Step 1 − Ontology Requirements Specification

The aim of step 1 of the methodology is, from use cases repository defined in WP1, to (*i*) analyse each use case of pilots, (*ii*) delimit the scope of the ontology, (*iii*) define the competency questions and, (*iv*) list the relevant terms. In the common template, each responsible of pilot briefly described the goal of the use case that will be designed, the main mission of the ontology design and proposed a set of questions that the ontology should be able to answer.

The use case of pilot #3a aims to design semantically a building, HVAC systems used in this building and its properties. As a result of the meetings with the business experts, different natural language questions are defined to validate the semantic data model such as:

- What are the zones of a building?
- What are sensors located in a zone?
- In which zone an HVAC valve is located?

In this step, a list of terms that are relevant for the domain of knowledge, are also extracted. The extraction of terms is done by identifying the list of nouns and verbs. Nouns represent the list of candidate terms that will be used as the preliminary concepts in ontology modules. This list of terms should be non-redundant and is used to validate the concept coverage after designing the ontology.

5.2   Application of Step 2 - Ontology Analysis

The goal of this step is to identify the relevant concepts needed to be modelled in the semantic data model and the reused or extended ontologies. Concretely, for each term or a group of terms extracted in the previous step presents in the table below, we associate a concept name or a relation name that will be used in semantic data model.

Then, if the concept already exists in an ontology, we put the existing concept in the column "existing ontology". In the case that the concept exists in several ontologies we put all the concepts in the same cell to prepare the mapping between these concepts. Otherwise, if the concept does not exist yet in ontologies, we seek for the concept that could be extended with our new concept and put it in the "Extending ontology" column. Reusing ontologies is not straightforward, as we can unfortunately be confronted with modelling errors such as the confusing between *is-a* and *part-of* relationship. This error is obviously serious because it corrupts the global consistency that leads to inference errors when querying the knowledge graph using these ontologies.

5.3   Application of Step 3 - Overview of Ontological Model

The purpose of this step is to harmonize all diagrams proposed for each pilot and provide an illustrative example for some semantic data models. The harmonization plays an important role in pilots that present different diagrams for each use case. Regarding the similarity or difference existing between these models, some effort is being made to address this issue; an example is the efforts focusing on the analysis of same concepts between models of pilots. The fact is that each responsible of pilot has defined his model, but he/she does not have a sufficient overview of the existing ontologies, so he/she created a new concept or relation instead of reusing an existing one. We hoped to have a tool that allow us to generate these similitude, but all this step is done manually. We also deal with all aspects related to the unified use of other concepts involved in a harmonization process, e.g., granularity, harmonization strategy, model quality, etc. This

section outlines the harmonization of multiple semantic data models that will
be bring some benefits such as a clear definition of concepts and relations used
in each pilot, a reference semantic data models under the same structure, with
uniform and formal vocabulary, etc.

5.4   Application of Step 4 - Formalization of Semantic Data Models

Formalization is the last step to do in the proposed methodology and it is done
after multiple interactions with stakeholders to validate that the models meet
the use cases needs. Thus, formalization consists of representing, thanks to an
ontology web language, the result of Step 3 which was illustrated in several
diagrams. Notice that, concepts related to a specific domain, are defined in the
same module. All resulted modules constitute the PLATOON ontology.

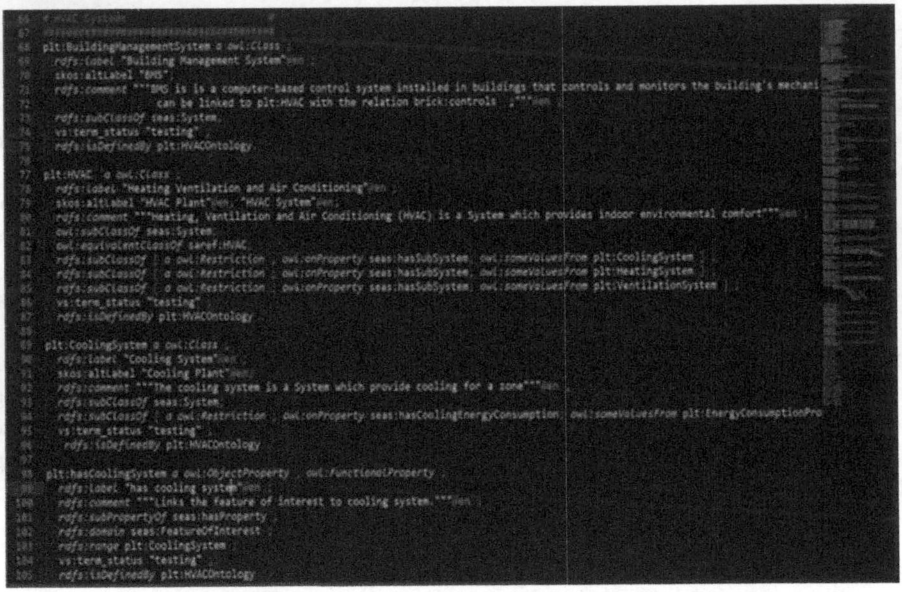

Fig. 3. An extract of the formalization of the HVAC module

The OWL2 DL language is used to describe each module, with choosing the
TURTLE serialization which is human and machine readable. Figure 3 shows an
extract of the formalization of the HVAC ontology module. Each new concept
is defined as `owl:Class` (see the line 77). Its label is defined with *rdfs:label*
annotation in the line 78. If the concept can have other labels, *skos:altLabel*
is used to define all the alternative labels (see the line 79). The definition associated to the concept is defined with the *rdfs:Comment* annotation in the
line 80. The subsumption relation is defined by the *rdfs:subClassOf* relationship. If there is an overlapping with other ontologies we use the relation

$owl:equivalentClassOf$ to associate to other existing concept like in line 82 where we put $plt:HVAC$ as equivalent to $saref:HVAC$. From the line 83 to 85, there are restrictions that are necessary properties to be an HVAC system. That means, every HVAC system has as subsystem a cooling system, a heating system and a ventilation system. The line 86, $vs:term\_status$ is an annotation to indicate if the concept is on testing phase or already validated. The line 87 is annotation to indicate which ontology module define this concept.

5.5   Use Case Instantiation with an Illustrative Example

This section aims to describe how the semantic data model (ontologies) can be used for a given dataset or use cases. Table 1 shows an example of dataset (raw data) that related to the temperature in a building. We need to represent semantically the temperature in the building which an important parameter to optimize and control the building occupancy. Five columns are defined:

- **BuildingID**: includes all the identifiers of buildings
- **ZoneID**: includes all the identifiers of zones
- **TempSensor**: includes all the names of sensors
- **Value C°**: includes all temperature values in degrees Celsius
- **Date**: includes all dates of temperature measurement.

Table 1. Datasets of temperature in a building

BuildingID	ZoneID	TempSensor	Value C°	Date
1	1	S1	22	20201008:11h40
1	1	S1	21	20201008:11h50
1	2	S2	20	20201008:11h40
1	2	S2	19	20201008:11h50
2	1	S1	17	20201008:7h00
2	1	S1	20	20201008:8h00

Figure 4 shows the result obtained by transforming raw data into semantic data. This example gives ideas for the data integration phase in how the semantic data model should be exploited.

The building is identified by an URI for example for the first building it is identified bu the URI `<engie/building/1>` and which has the type `s4bldg:Building` (see orange box). This building contains ($bot:containsZone$) different zones identified as follows `<engie/building/1/zone/1>` and `<engie/building/1/zone/2>`. Each zone is defined thanks to three relevant relations (see green box):(i) is−a: in order to determine its type as `bot:Zone`; (ii) $seas:temperature$: in order to link it to its temperature `<engie/building/1/zone/1/airtemperature/property>`;

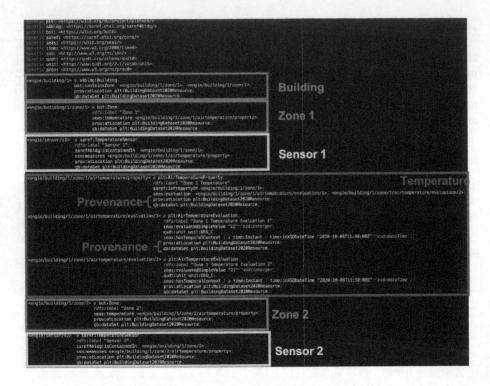

Fig. 4. Extract of Temperature in building: RDF description (TTL file

(*iii*) *prov:atLocation* : in order to specify the provenance of the data. A sensor `<engie/sensor/s1>` is a `saref:TemperatureSensor` (see yellow box). It is contained in (*s4bldg:isContainedIn*) in a zone `<engie/building/1/zone/1>` and measures (*ssn:measures*) the air temperature `<engie/building/1/zone/1/airtemperature-/property>`.

The temperature `<engie/building/1/zone/1/airtemperature/property>` *is-a* `plt:AirTemperatureProperty` (see pink box). It is a property of (*saref:isPropertyOf*) this zone `<engie/building/1/zone/1>`.

6   Discussion

The aim of this article, is to give a feedback about the experiences we had in the PLATOON H2020 project, and to share the methodology followed with the results obtained. As presented in this paper, we had 19 use cases in seven pilots. The uses cases were transversal and encompass several domains, which lead us to create several ontological modules. As a first step before starting to defines the PLATOON semantic data model, we made an inventory of the ontologies and data model that could be candidate to cover some the defined use cases. Saref and CIM considered as a standard in the Energy community was a good candidate to be reused in the PLATOON project. However, The result of our inventory is twice: i) there are some overlapping between several existing ontologies, then for interoperability issues, it is necessary to do mapping between similar concepts/relations, and ii) all theses existing ontologies taken together can not meet the need of the uses cases defined in PLATOON pilots. Then Creating a Semantic data model become necessary.

In order to organize the work with the different participants and stakeholders, we schedules weekly meetings to work with the business experts concerning each pilot and bi-weekly meeting with all teams to follow the progress of all pilots. In fact, the interactions was done earlier in the process by working with business experts that are in the same time stakeholders. During this work, we faced, some difficulties, we can cite the: (i) choosing between existing concepts, or relations but that not wholly feet the conceptualization needs or the as noticed previously confusion in the subsumption relation (is-a) like in brick and CIM; furthermore CIM does not use semantic relations between concepts but it is simple link based on the name of two linked concept. In fact, CIM is resulted on the translation of an UML models into OWL. The in our point of view it is not build with ontological paradigm; (ii) misunderstanding of some business experts about the philosophical paradigm of ontologies and the importance to create the better taxonomical structure with eliminating the subsumption confusion. (iii) Concepts definition are also not obvious, most of the time they oriented functional properties not "being" properties.

The ontological data model was build in two phases: the first phase, was based on the LLUC and the competency questions defined in the step 1, and the second phase, was based on the reel data that need to be transformed in semantic data to be integrated in the pilots validation process. As expected, some missing concepts and relations were defined in this second phase. Each partner developed a semantic data pipeline to ingest the data based on the semantic data model, however this transformation part is not detailed in this paper. Finally, it is important to notice that PLATOON ontologies is use case oriented and do not pretend that cover all the energy system domain.

7   Conclusion

In this paper, we present the methodology used to build a common semantic data models for 19 LLUCs of 7 Pilots. These LLUCs fully address PLATOON's objectives and challenges. Each pilot identifies a set of datasets and data exchange requirements which served as the basis for the construction of the common semantic data models, as described in the Sect. 2. From the descriptions of LLUCs, we identified five topics that covers the pilot's needs. These pilots shared a set of common-sense concepts related to different domains in particular Building and HVAC domain (pilots #3a, #3b and #3c), Renewable energy domain (pilot #1a) and Grids domain (pilots #2a, #2b and #4a). All pilots used also similar notions about the weather, the measurement, forecasting, failures, etc. To achieve a richer interoperability between pilots through the handling of data heterogeneity, we propose a specific methodology to create harmonized semantic data models that include all the needs of the PLATOON project's use cases. This paper puts forward the construction of common semantic data models by reusing existing domain ontologies (SAREF, CIM, SEAS, etc.), as an essential part, or/and constructing ontology modules from the beginning of the ontology design process. We have elicited four major steps required for performing semantic interoperability, after examining the existing ontologies of the energy sector. The step 1 called "ontology requirements specification", allows use cases analysis, ontology scope definition, terms elicitation and competency questions determination. Step 2 named "ontology analysis", aims at analysing the competency questions, and identifying ontology modules by reusing, extending and/or creating an ontology. Step 3 "overview of ontological modules", is about putting all the modules together, evaluating ontologies and instantiating some use cases with an illustrative example. The last step concerns the discussion with stakeholders and the formalization process of the common semantic data models. All four steps are intertwined, and each step provides the necessary input to perform the next step. We applied the steps of the proposed methodology in different LLUCs of pilots. We presented some fragments of semantic data models such as Building, HVAC, Wind Turbine, Grid, etc. Finally, we presented an example of formalization of the semantic data model thanks to the ontology web language; and illustrated an example of semantic representation that transform the data in knowledge by using the semantic data model created in the PLATOON project during the two phases. The output of the task is to provide an OWL file for each new ontological module and an RDF file for each use case. As a perspective on the results of this work, we can of course cite the use of these ontological SDM modules in the PLATOON project, but the results will also be used as base modules in the ongoing initiatives within GAIA-X[1] and European Data Space such as Enrshare[2] Project and OMEGA-X[3] project where we are participating as a task leader.

Acknowledgement. The research presented in this paper is partly financed by the European Union (H2020 PLATOON, Pr. No: 872592; H2020 LAMBDA, Pr. No: 809965), and partly by Engie.

References

1. Contreras, M.C.B., Reyes, L.F.H., Ortiz, J.A.R.: Methodology for ontology design and construction. Contaduría y Administración (2019)
2. Gómez-Pérez, A., Fernández-López, M., De Vincente, A.: Towards a method to conceptualize domain ontologies. In: ECAI-96 Workshop on Ontological Engineering (1996)

---

[1] https://gaia-x.eu/.

[2] https://enershare.eu/.

[3] https://omega-x.eu/.

3. Grüninger, M., Fox, M.: The design and evaluation of ontologies for enterprise engineering (1995)
4. Hitzler, P., Krisnadhi, A.: A tutorial on modular ontology modeling with ontology design patterns: the cooking recipes ontology (2018)
5. Lenat, D.B., Guha, R.V.: Building Large Knowledge-Based Systems; Representation and Inference in the Cyc Project. Addison-Wesley Longman Publishing Co., Inc., Boston (1989)
6. McMorran, A.W.: An introduction to IEC 61970–301 & 61968–11 : The common information model (2007)
7. Schreiber, A.T., Wielinga, B.J., Jansweijer, W.N.H., Anjewierden, A., van Harmelen, F.: The kactus view on the 'o' word. In: IJCAI 1995 (1995)
8. Sure, Y., Staab, S., Studer, R.: Methodology for development and employment of ontology based knowledge management applications. SIGMOD Rec. **31**, 18–23 (2002)
9. Swartout, B., Patil, R., Knight, K., Russ, T.: Toward distributed use of large-scale ontologies. in: ontological engineering. In: AAAI-97 Spring Symposium Series, pp. 138–148 (1997)
10. Uschold, M., Gruninger, M.: Ontologies: principles, methods and applications. Knowl. Eng. Rev. **11**, 93–136 (1996)

Semantic Carbon Footprint of Food Supply Chain Management

Swapnali Yadav[1], Megan Powers[1], Edlira Vakaj[1], Sanju Tiwari[2],
and Fernando Ortiz-Rodriguez[2(✉)]

[1] Birmingham City University, Birmingham, UK
[2] Universidad Autonoma de Tamaulipas, Ciudad Victoria, Mexico
ferortiz@uat.edu.mx

Abstract. Currently climate change poses a significant challenge to us, both now and as we head into the future. Several individuals endeavour to adopt more sustainable lifestyles, ensuring that our daily choices do not have adverse impact on our planet. Under the umbrella of sustainability, food stands out as an area of specific emphasis, specifically regarding the environmental aspects of our selected diet and the ways we can make decisions that reduce the negative impacts. It is challenging to figure out how to effectively model the environmental impact of food production in a way that is useful to food manufacturers, consumers, and other individuals concerned about sustainability. To tackle this challenge, an ontology and a knowledge graph have been developed to assess the CO_2 emissions generated throughout the several stages of food manufacturing and distribution. This comprehensive approach leads as a significant indicator of environmental impact, explores potential alternatives, and provides a comparison of CO_2 emissions. Our work focuses specifically on assessing the environmental consequences of food production, primarily through the measurement of CO_2 emissions. We consider key factors such as animal feeding, land usage, and transportation within this scope. The data to test our approach is coming from various open-source datasets such as the Food Emissions Dataset, Nutrition Facts Dataset, etc. Several use-cases have been simulated to validate the usability and efficiency of the work.

Keywords: carbon footprint · food supply chain · ontologies

1 Introduction

Climate change is one of the most significant challenges we face, both now and as we head into the future. Several individuals are dedicated to adopting more sustainable lifestyles, aiming to ensure that our daily decisions have no detrimental repercussions on our planet. A specific attention has been directed towards food as it is one of the prominent areas of sustainability. It is a big challenge to effectively model the environmental consequences of food production to be significant to food manufacturers, consumers, and other stakeholders focused on sustainability. To address these challenges, we have proposed an ontology (Yadav et al. 2023; Vakaj et al. 2023) concentrated on CO_2 emitted by the different stages involved in manufacturing and distributing food as a barometer for potential alternatives, environmental impact, and how the CO_2 emissions scale up.

© The Author(s), under exclusive license to Springer Nature Switzerland AG 2023
S. Tiwari et al. (Eds.): AI4S 2023, CCIS 1907, pp. 202–216, 2023.
https://doi.org/10.1007/978-3-031-47997-7_16

Other researchers have contributed to this as well such as Babaie et al. (2019) modelled FEWsOnt ontology with significant structural and dynamic concepts such as security, footprint, challenge, risk, impact and uncertainty of food-energy-water (FEW) systems using naturalistic and scheduled processes. In addition, ontology aids participants not only to retrieve new inferences with the help of semantic visions but also to find insights and associations of systems' components for enhanced decision making. The users can optimize or control the unfavourable changes of FEW systems regarding natural and social systems.

At present, many food systems are unsustainable causing significant resource depletion and inappropriate environmental effects. This issue is severe as it is argued that today's food consumption is equivalent to a fossil resource. The transition to sustainable food systems will provide a chance to use global resources more efficiently to develop resilience systems and facilitate governance regarding communication education and computation (Holden et al., 2018). 'Internet of Food' represented the idea of food items containing an 'IP identity', which introduced the query of how this might affect our eating practices (Foley et al., 2011). It focused on technology that could influence food choices and predicted scrutinizing food entities remotely through the Internet.

In 2018, Craig launched an ontology called FoodOn to describe food for human, household and agricultural animal consumption. It is based on LanguaL™, which is a framework for Food Description Thesaurus. It is expanding its class structure to differentiate between single-source and multi-source foods and developing basic models of food product representation, diet, nutrition, and additive. Adding to the state of the art we capture the whole supply chain CO_2 emissions life cycle of a specific food making use of open data.

Rest of the sections are as follows: Sect. 2 has discussed Ontology Methodology, Sect. 3 has presented the Ontology Design with ontology requirements, competency questions and evaluation with SPARQL queries. Finally concluded the paper in Sect. 4.

2 Ontology Methodology

The scope of the proposed ontology encompasses the environmental consequences of food production as measured by CO_2 emissions and categories including land use, transportation, and feeding animals. The proposed ontology is constrained to this specific scope to avoid ontology creation too broad to be intuitive or understandable for stakeholders. As a constraint, the proposed work will not cover a narrow part of the food production tasks such as transportation, because this would mean that a holistic picture of carbon emissions cannot be created.

For instance, wheat production may exhibit huge emissions linked with transportation, but there is no emissions when it comes to animal feed. The proposed ontology has been designed without applying any existing ontologies in the construction of this project.

2.1 Ontology Requirements Specification

1. The ontology shall record information about CO2 emissions created by different stages of the food manufacturing process.
   a. The ontology shall record information about CO2 emissions from feeding animals.
   b. The ontology shall record information about CO2 emissions from farming.
   c. The ontology shall record information about CO2 emissions from using land resources.
   d. The ontology shall record information about CO2 emissions created by selling products at retail outlets.
   e. The ontology shall record information about CO2 emissions from packaging products.
   f. The ontology shall record information about CO2 emissions from transporting products.
2. The ontology shall record the amount of CO2 associated with individual instances of general classes.
3. The ontology shall include a variety of individual instances of generic classes (e.g. a Gala Apple from a specific country of origin being an instance of an Apple class).
4. The ontology shall use the specific instances of generic classes to compare and contrast the CO2 emissions from different related instances.
5. The ontology shall represent different recipes that are made from ingredients represented in the ontology.
6. The ontology shall present the CO2 emissions associated with recipes modelled in the ontology.
7. The ontology shall contain nutrition information associated with individual food instances in order to allow users of the ontology to make substitutions that are both environmentally conscious and have similar nutrition content to foods that have a larger environmental impact.
   a. The ontology shall record the protein content of different food instances.
   b. The ontology shall record the carbohydrate content of different food instances.
   c. The ontology shall record the fat content of different food instances.
8. The ontology shall record whether a food is high or low in carbohydrates, fat, and/or protein.
9. The ontology shall calculate the amount of emissions from a hypothetical grocery basket that consists of various instances of food.
10. The ontology shall record the country of origin for specific instances of food.
11. The ontology shall categorize instances of items as high emission, medium emission, or low emission following SWRL rules.
12. The ontology shall calculate and present equivalent emissions in terms of electricity usage for a given specific food instance

2.2 Competency Questions

1. Query for nutritious food items with high protein but with low carbon emission.
2. Query for high emission food items with their substitutes.
3. Query for Grocery bag with a category of Grain and Milk with low emission.
4. Query for equivalent emissions for food items.
5. Query the ontology to retrieve the maximum producer of CO2 emissions from all of the food items with the category "foodEmissionsProcessing".
6. Query the ontology to retrieve the average carbohydrate levels of food items that have more than 50 g of carbohydrates.
7. Query the ontology to retrieve all instances with "Wine" in the name and their total emissions.
8. Query the ontology to retrieve the total CO2 emission from "Rice" and with CO2 less than rice
9. Querying for all grains whose total emissions are less than the total emissions of rice, using the:foodTotalEmissions data property.
10. Query to find the highest emission from poultry.
11. Query to find all poultry animals with CO2 emission (emission other than the highest emission from:PoultryDuck)
12. User queries to retrieve all recipes containing a certain food product. She wants to use a general keyword because the user is unsure of more specific categories.
13. A user wants to know more about substitutions for specific recipes and queries recipes that contain dairy in them.

2.3 Users

Below actors, who want to use the ontology for different purposes have been shown in order to represent the intended users of the ontology.

1. User A
   a. Loretta is a conscious consumer who wants to learn about the impact her food choices have on the environment. She is interested in learning more about the emissions of food in dishes she commonly cooks, as well as good substitutes for food she likes.
2. User B
   a. Malcolm is an employee at Crossroads Food Co., a food company based in the United Kingdom. He is involved in several sustainability initiatives and wants to use the ontology to help gauge the carbon emissions of foods that the company deals in.
3. User C
   a. Nadia is a farm owner living in the United Kingdom who mainly produces wheat and barley. She is aware that consumers want to explore more sustainable options, so she is interested in expanding her business to produce other food products. She wants to make use of the ontology to see similar food products that she can invest in.

4. User D
   a. Thomas is a university student who has just begun to consider alternatives to the food that he eats. Because he has not paid much attention to factors such as nutrition or emissions, he doesn't know very much about what to look for when it comes to sustainable alternative options. He wants to use ontology to explore information about the food he eats in a way he can understand.
5. User D
   a. Maria, an agricultural officer as well as an environmentalist wants to calculate the amount of CO_2 produced from the rice paddy field in her village. She also wants to check the CO_2 emission produced from other grains whose total emissions are less than the total emissions of rice. She wants to use this ontology for resolving her query.
6. User E
   a. The Dudley Poultry Council decided to check the CO_2 released from poultry animals and poultry which release the highest amount of CO_2. The local body wants to use this ontology to solve their query.

2.4 Intended Use

The primary usage is to function as a reliable resource for individuals keen on monitoring the CO_2 emissions stemming from food production. The designed ontology aimed to provide precise information about CO_2 emissions from food manufacturers and alternatives to high-emission food products for business owners, industry workers, consumers, and other stakeholders who are engaged in issues relating to sustainability. These stakeholders will be able to use the ontology features as a point of reference to inform their respective requirements.

The ontology is designed to illustrated the environmental significance, broken down by processing stage, of several common food items. It is also able to present alternative food items, the total environmental impact of dishes, and equivalent emissions measured in terms of electricity emissions. In order to describe the potential applications by various different types of actors, use cases are described in the 'Use Cases' section.

3 Ontology Design

In ontology, there are a total of 74 classes (`Fruits`, `Food`, `Barley`, `Apple`, `Wine`, `Drink`, `TikkaMasala`, `Oil`, `Recipe`, `UnitedKingdom`, `Country`, `Brazil`, `EmissionsLevel`, `ElectricalAppliance`, etc.); 10 Object properties (`hasEquivalentEmission`, `hasIngredient`, `hasEmissionLevel`, `hasItem`, `hasLessEmissionsThan`, `hasOrigin`, `hasMoreEmissionsThan`, `isIngredientIn`, `isOriginOf`); 12 Data properties (`fatLevel`, `carbohydrateLevel`, `foodEmissionsAnimalFeed`, `electricalEmissions`, `foodEmissionsLandUse`, `foodEmissionsFarm`, `foodEmissionsProcessing`, `foodEmissionsPackaging`, `foodEmissionsRetail`, `foodEmissionsTotal`, `foodEmissionsTransport`, `proteinLevel`); 76 Individuals (`ApplesGala`, `BeefIrishMeat`, `CentralAirConditioner`, `ApplesGrannySmith`, `Seitan`, `WineLouisRoedererChampagne`, `RapeseedOil` etc.).

3.1 Data Sources

See Table 1.

Table 1. Data sources with corresponding details

Data	Characteristics	Description	Owner	Source
Food Emissions Dataset	Data Spreadsheet	A dataset by Our World in Data detailing food emissions data	Our World in Data	https://tinyurl.com/food-emissions-data-set
Nutrition Facts Dataset	Search Tool	A dataset by USDA Food Data Central detailing the nutrition facts of various food items	USDA Food Data Central	FoodData Central (usda.gov)
Dishes Dataset	Report	A report by the WWF that contains information about the food and emissions in different dishes	World Wildlife Federation (WWF)	Food_in_a_warming_world_report.PDF (wwf.org.uk)
Equivalent Electrical Equipment Emissions Graph	Graph	A graph by Food Navigator showing appliance emissions compared to food	Food Navigator	Consumers underestimate the emissions associated with food but are aided by labels	Nature Climate Change

3.2   Evaluation

3.2.1   SPARQL Queries for Competency Questions

- Prefixes:

```
PREFIX :
<http://www.semanticweb.org/swapnali/ontologies/2021/11/CarbonFootprin
tOfFoodSupplyChain#>
PREFIX owl: <http://www.w3.org/2002/07/owl#>
PREFIX rdf: <http://www.w3.org/1999/02/22-rdf-syntax-ns#>
PREFIX rdfs: <http://www.w3.org/2000/01/rdf-schema#>
```

1. Query for nutritious food items with high protein but with low carbon emission

```
SELECT   ?food   (STR(?lab)   AS   ?label)   ?proteins   ?carbs   ?fats
?carbonEmission WHERE {
           ?food rdf:type :HighProteinFood;
           rdf:type :LowCarbohydrateFood;
           rdf:type :LowFatFood;
           :hasEmissionLevel :LowEmission;
           :foodEmissionsTotal
           ?carbonEmission;
           :proteinLevel
           ?proteins;
           :carbohydrateLevel
           ?carbs;
          :fatLevel ?fats.
          OPTIONAL {?food rdfs:label ?lab}
```

2. Query for high emissioned food items with its substitutes.

```
SELECT   ?food   (STR(?lab)   AS   ?label)   ?substitutes
WHERE {
           ?food :hasEmissionLevel :HighEmission;
           :hasSubstitute ?substitute .
           ?substitute rdfs:label ?substitutes
           OPTIONAL {?food rdfs:label ?lab}
}
     ORDER BY ASC(?food)
```

3. Query for Grocery bag with a category of Grain and Milk with low emission.

```
SELECT ?x (SAMPLE(?y) AS ?food) (SAMPLE(?lab) AS ?label)
    (MIN(?emissions) AS ?minEmissions) WHERE {
                ?x rdfs:subClassOf :Food.
                ?y rdfs:subClassOf ?x;
                :hasEmissionLevel :LowEmission;
                :foodEmissionsTotal ?emissions.
                ?y rdfs:label ?lab
                FILTER (LANG(?lab) = "en")
                FILTER (REGEX(STR(?x), "Grain") || (REGEX(STR(?x),
"Milk")))
    }
    GROUP BY ?x
    ORDER BY ?x
```

4. Query for equivalent emissions for food items

```
SELECT        ?food      ?foodEmission      ?electricalAppliance
?electricalEmission
    WHERE {
    ?food :hasEquivalentEmission ?electricalAppliance;
    :foodEmissionsTotal ?foodEmission .
    ?electricalAppliance :electricalEmissions ?electricalEmission
    }
    ORDER BY ?electricalAppliance
```

5. Query the ontology to retrieve the maximum producer of CO2 emissions from all of the food items with the category "foodEmissionsProcessing".

```
SELECT (MAX(?z) as ?maxEmission) WHERE {
                ?e rdfs:subClassOf :Food .
                ?x rdfs:subClassOf ?e .
                ?y rdf:type ?x .
                ?y :foodEmissionsProcessing ?z}
```

6. Query the ontology to retrieve the average carbohydrate levels of food items that have more than 50 g of carbohydrates.

```
SELECT (AVG(?carbs) as ?averagecarbs) WHERE {
                ?g rdfs:subClassOf :Grain .
                ?food rdf:type ?g .
                ?food :carbohydrateLevel ?carbs .
                FILTER (?carbs > 50)
```

7. Query the ontology to retrieve all instances with "Wine" in the name and their total emissions, without using the: foodTotalEmissions property.

```
SELECT (SAMPLE(?food) as ?f) (SUM(?fepr + ?fepa + ?fean+
?fefa+
    ?fela+ ?feta) as ?totalEmissions) WHERE {
               ?food rdf:type :Wine .
               ?food :foodEmissionsProcessing ?fepr .
               ?food :foodEmissionsPackaging ?fepa .
               ?food :foodEmissionsAnimalFeed ?fean .
               ?food :foodEmissionsFarm ?fefa .
               ?food :foodEmissionsLandUse ?fela .
               ?food :foodEmissionsTransport ?feta . }
    GROUP BY (?food)
```

8. Query to find Total CO2 Emissions of rice

```
    SELECT   (SAMPLE(?food)  as  ?f)  (SUM(?fepr  +  ?fepa  +
?fean+ ?fefa+
    ?fela+ ?feta) as ?totalEmissions) WHERE {
               ?food rdf:type :Rice .
               ?food :foodEmissionsProcessing ?fepr .
               ?food :foodEmissionsPackaging ?fepa .
               ?food :foodEmissionsAnimalFeed ?fean .
               ?food :foodEmissionsFarm ?fefa .
               ?food :foodEmissionsLandUse ?fela .
               ?food :foodEmissionsTransport ?feta . }
    GROUP BY (?food)
```

9. Querying for all grains whose total emissions are less than the total emissions of rice, using the: foodTotalEmissions data property.

```
    SELECT ?grain ?y WHERE{
               ?x rdfs:subClassOf :Grain .
               ?grain rdf:type ?x .
               ?grain :foodEmissionsTotal ?y .
               ?r rdf:type :Rice .
               ?r :foodEmissionsTotal ?re
               FILTER (?y < ?re)
    }
```

10. Query to find the highest emission from poultry.

```
Select ?food ?emissions where {?food rdf:type :Poultry . ?food
:foodEmissionsTotal ?emissions }
```

11. Query to find all poultry animal with CO2 emission (emission other than highest emission from:PoultryDuck)

```
SELECT ?z ?y WHERE{
            ?x rdfs:subClassOf :Meat .
            ?z rdf:type ?x .
            ?z :foodEmissionsTotal ?y .
            ?r rdf:type :Poultry .
            ?r :foodEmissionsTotal ?re
            FILTER (?y < ?re)
}.
```

12. Loretta queries to retrieve all recipes containing a certain food product. She wants to use a general keyword because she's unsure of more specific categories.

```
SELECT ?recipe ?y WHERE { ?recipe rdfs:subClassOf :Recipe;
 :hasIngredient ?y
FILTER regex(str(?y), "Eggs") }
```

13. Loretta, wanting to know more about substitutions for specific recipes, queries recipes that contain dairy in them.

```
SELECT DISTINCT(?y AS ?meal) ?x WHERE{
?x :isIngredientIn ?y .
?x rdf:type :Dairy .
FILTER (regex(str(?x), "Cheese") || regex(str(?x), "Milk"))
```

3.2.2 Use Cases

1. *Use Case 1*

This use case is intended to satisfy the requirement of representing an average consumer's query about the items that they buy regularly.

Loretta wants to check the hypothetical result of putting several items she regularly buys to see their total combined CO2 emissions, as well as what the highest contributing categories to CO_2 emissions are. Loretta pulls up the CO_2 Emissions Ontology. She queries several items and then orders them to see which ones have the highest total emissions. She finds out that the beef she has been buying is a high contributor to emissions, so she queries again to find other items that are high in protein. She notices that poultry, setain, and tofu are high in protein and are lower in emissions.

She wants to use the semantic model to query information about the CO_2 emissions of a hypothetical trip to the grocery store. She also wants to see the highest contributors to emissions, and what good substitutions for them are.

2. Use Case 2

This use case is intended to satisfy the needs of a corporate worker who is interested in using the ontology to measure the carbon emissions of food that his company trades in, as well as potential alternatives that are more sustainable.

Malcolm wants to check various food products that his company sells in order to see what their emissions are. He also wants to check what the equivalent emissions for the food products are. Malcolm pulls up the CO_2 Emissions Ontology. He queries what food products are in the category "high emission". He also queries the different categories of emissions per food item to see what parts of the manufacturing process have the highest emissions. He then queries "low emission" food items. Finally, he queries the equivalent electricity emissions for high emission food items.

He wants to use the semantic model to query information about which food items are high and low emissions. He also wants to know about a point of comparison for the number of emissions from food items in terms of electricity emissions.

3. Use Case 3

This use case is intended to satisfy the requirement of representing a business owner's usage of the ontology to research potential new products to sell.

Nadia wants to see the result of using the ontology to query whether the food items she sells are high-emission. She uses the ontology to check the country of origin, to see if there are food items that are more commonly grown in the United Kingdom that she can sell. She uses the ontology to check the nutrition of what she sells, as well as alternatives with similar levels of nutrition.

She wants to use the semantic model to query information about the food products she sells. She wants to check the nutrition of the food products she sells. She also wants to see which food items are presented as good substitutes for popular food products.

4. Use Case 4

This use case is intended to satisfy the requirement of representing someone who is not very aware of matters related to sustainability but wants to use the ontology to learn more.

Thomas wants to use the ontology to find out more about the breakdown of emissions by food product. He uses the ontology to query each item he normally purchases one by one to see the breakdown by emissions. After this, he begins querying nutrition facts of products with high emissions. Then, he queries for appliances that have similar electrical emissions to food. Finally, he queries for substitutions.

He wants to learn more about the breakdown of emissions related to the manufacture of food items that he commonly buys. He wants to understand more about the nutrition facts of those food items, as well as substitutions he can make based on nutrition. Finally, he wants to find out what electrical appliances have similar emissions to food items that he buys.

5. *Use Case 5*

This use case is intended to query the C02 emission of rice paddy fields in an area and query some grains whose total emissions are less than the total emissions of rice.

Marie, an agricultural officer wants to check the CO_2 emission level of rice in the paddy field in her village. Maria pulls up the CO_2 Emissions Ontology. She wants to know the total emission of CO_2 from rice and then she checks grains whose total emissions are less than the total emissions of rice. So the less CO_2 emitting crop or grain can be cultivated in the field, to reduce CO_2 emission.

She found that rice is with total emission of 3.9. She sees that rice has the highest emission when compared with Barley, Maize, Wheat, and Oatmeal with 1.1, 1.1, 1.4 and 1.6 of CO_2 emissions respectively. Hence, she decided better to cultivate wheat, maize, or oatmeal to reduce CO_2 emissions in her area.

6. *Use Case 6*

This use case is intended to query the C02 released from poultry and query poultry with the highest CO_2 emission.

Dudley Poultry decided to reduce the CO_2 emission from the poultry in Dudley, they decided to check the highest amount of CO_2 released by poultry and the total emission from other poultry to decrease the CO_2.

They found that PoultryDuck is with the highest emission of 6.1 kg per tonne. They also found FishCodFarmed with 5.1 CO_2 emissions. A better understanding of the Carbon footprint of poultry will help to reduce CO_2.

3.2.3 SWRL Rules

See Table 2.

3.2.4 Performance Enhancement Tools

1. Create axioms from Excel workbook (Cellfie Plugin):

Protege provides a tool for importing spreadsheet data into OWL ontologies. The mentioned tool is to import the initial dataset in the form of a CSV file to Protege with the help of rules.

2. WebProtégé:

WebProtégé is a cloud-based application which allows users to collaboratively edit OWL ontologies and apply External edits from Local ontologies. It is used for project collaboration over the internet.

Table 2. SWRL Rules

Name	Description	Owner
TotalEmission	A rule to add emission values of each food supply chain stage	CarbonFootprintOfFoodSupplyChain:foodEmissionsAnimalFeed(?x, ?af) ^ CarbonFootprintOfFoodSupplyChain:foodEmissionsFarm(?x, ?f) ^ CarbonFootprintOfFoodSupplyChain:foodEmissionsLandUse(?x, ?lu) ^ CarbonFootprintOfFoodSupplyChain:foodEmissionsPackaging(?x, ?p) ^ CarbonFootprintOfFoodSupplyChain:foodEmissionsProcessing(?x, ?pr) ^ CarbonFootprintOfFoodSupplyChain:foodEmissionsRetail(?x, ?r) ^ CarbonFootprintOfFoodSupplyChain:foodEmissionsTransport(?x, ?t) ^ swrlb:add(?totalEmission, ?af, ?f, ?lu, ?p, ?pr, ?r, ?t) ^ swrlb:round(?roundedTotalEmission, ?totalEmission) -> CarbonFootprintOfFoodSupplyChain:foodEmissionsTotal(?x, ?roundedTotalEmission)
EquivalentEmission	A rule to assign equivalent electric appliance to food item.	CarbonFootprintOfFoodSupplyChain:foodEmissionsTotal(?x, ?a) ^ CarbonFootprintOfFoodSupplyChain:electricalEmissions(?z, ?b) ^ swrlb:equal(?a, ?b) -> CarbonFootprintOfFoodSupplyChain:hasEquivalentEmission(?x, ?z)
HighEmissionLevel	A rule to categorize food items to high emission level.	CarbonFootprintOfFoodSupplyChain:foodEmissionsTotal(?x, ?a) ^ swrlb:greaterThanOrEqual(?a, 50.0) -> CarbonFootprintOfFoodSupplyChain:hasEmissionLevel(?x, CarbonFootprintOfFoodSupplyChain:HighEmission)
MediumEmissionLevel	A rule to categorize food items to medium emission level.	CarbonFootprintOfFoodSupplyChain:foodEmissionsTotal(?x, ?a) ^ swrlb:greaterThanOrEqual(?a, 20.0) ^ swrlb:lessThan(?a, 50.0) -> CarbonFootprintOfFoodSupplyChain:hasEmissionLevel(?x, CarbonFootprintOfFoodSupplyChain:MediumEmission)
LowEmissionLevel	A rule to categorize food items to low emission level.	CarbonFootprintOfFoodSupplyChain:foodEmissionsTotal(?x, ?a) ^ swrlb:lessThan(?a, 20.0) -> CarbonFootprintOfFoodSupplyChain:hasEmissionLevel(?x, CarbonFootprintOfFoodSupplyChain:LowEmission)

4   Conclusion

In summary, an ontology has been developed to depict the realm of environmental impact arising from food production as measured through CO2 emissions. The ontology development process went through various stages, as it is required to figure out how to properly model our domain in a way that was not so broad that we were unable to complete our project, but not so narrow that only specific information could be queried.

During ontology development, it is found that anchoring it around food and the numerous types of CO2 emissions given off by several stages of the development process was significant. Furthermore, we explored that incorporating nutritional details alongside emissions proved valuable, serving as a benchmark for identifying comparable food items. Additionally, the inclusion of the country of origin was important, providing a basis for contrasting emissions linked with food products originating from different countries.

It was also useful in displaying the environmental cost of different common meals that used ingredients that were sourced from many different countries. This demonstrated how globalized the food we eat is, and how transportation emissions can often be a large point of contrast between otherwise similar food products.

Finally, we found that including equivalent electrical emissions as modelled through appliances to be a useful point of comparison. In the development of this project, we aimed to consider the needs of a variety of uses, not all of whom might understand the different categories of emissions or otherwise require a point of reference. This was a helpful lens to develop the project, as we had to keep the end-user in mind throughout.

Moving forward, there are some additions that could be made to the ontology. A larger sampling of food product instances would be useful to include in order to serve as a point of comparison. In addition, more classes and instances could be included representing other vegetarian and vegan protein alternatives because the process of producing meat products was found to be high in CO_2 emissions. Finally, we could expand the nutritional information associated with each food item to be more inclusive of factors such as fibre, protein, gluten content, and other such information. Nevertheless, we produced an ontology that offers a holistic, useful picture of the CO_2 emissions of food production.

References

Ottery, C.: Carbon emissions by local authority. The Guardian (2010). https://www.theguardian.com/news/datablog/2010/jul/08/carbon-emissions-local-authority. Accessed 17 May 2022

Richards, J.: Food's carbon footprint. Green Eatz (2016). https://www.greeneatz.com/foods-carbon-footprint.html

Ritchie, H.: You want to reduce the carbon footprint of your food? Focus on what you eat, not whether your food is local. Our World in Data (2020). https://ourworldindata.org/food-choice-vs-eating-local

foodnavigator.com (n.d.).: Study confirms carbon label efficacy: 'they had the predicted effect... lower-emission food choices'. https://www.foodnavigator.com/. https://www.foodnavigator.com/Article/2018/12/18/Carbon-labels-drive-more-sustainable-food-choices-scientists-confirm

Food in a Warming World the Changing Foods on the British Plate. (n.d.). https://www.wwf.org.uk/sites/default/files/2018-03/Food_in_a_warming_world_report.PDF

Babaie, H., Davarpanah, A., Dhakal, N.: Projecting pathways to food–energy–water systems sustainability through ontology. Environ. Eng. Sci. **36**(7), 808–819 (2019). https://doi.org/10.1089/ees.2018.0551

Holden, N.M., White, E.P., Lange, Matthew. C., Oldfield, T.L.: Review of the sustainability of food systems and transition using the Internet of Food. NPJ Sci. Food **2**(1), 18 (2018).https://doi.org/10.1038/s41538-018-0027-3

Foley, J., et al.: Solutions for a cultivated planet. Nature **478**, 337–342 (2011). https://doi.org/10.1038/nature10452

Craig, H.: Agricultural ontologies in use: FoodOn – a farm to fork food description ontology. CGIAR Platform for Big Data in Agriculture (2018). https://bigdata.cgiar.org/blog-post/agricultural-ontologies-in-use-foodon-a-farm-to-fork-food-description-ontology/

Kendall, E.F, McGuinness, D.L.: Ontology engineering. Synth Lect. Semant. Web Theory Technol. **9**(1), i-102 (2019)

Vakaj, E., Tiwari, S., Mihindukulasooriya, N., Ortiz-Rodríguez, F., Mcgranaghan, R.: NLP4KGC: natural language processing for knowledge graph construction. In: Companion Proceedings of the ACM Web Conference 2023, p. 1111 (2023)

Yadav, S., Powers, M., Vakaj, E., Tiwari, S., Ortiz-Rodriguez, F., Martinez-Rodriguez, J.L.: Semantic based carbon footprint of food supply chain management. In: Proceedings of the 24th Annual International Conference on Digital Government Research, pp. 657–659 (2023)

Author Index

S. Tiwari et al. (Eds.): AI4S 2023, CCIS 1907, pp. 217–218, 2023.
https://doi.org/10.1007/978-3-031-47997-7

Printed in the United States
by Baker & Taylor Publisher Services